REPTILES
AND
AMPHIBIANS
Care • Behavior • Reproduction

Elke Zimmermann
Translated by U. Erich Friese

Distributed in the UNITED STATES to the Pet Trade by T.F.H. Publications, Inc., One T.F.H. Plaza, Neptune City, NJ 07753; distributed in the UNITED STATES to the Bookstore and Library Trade by National Book Network, Inc. 4720 Boston Way, Lanham MD 20706; in CANADA to the Pet Trade by H & L Pet Supplies Inc., 27 Kingston Crescent, Kitchener, Ontario N2B 2T6; Rolf C. Hagen Ltd., 3225 Sartelon Street, Montreal 382 Quebec; in CANADA to the Book Trade by Macmillan of Canada (A Division of Canada Publishing Corporation), 164 Commander Boulevard, Agincourt, Ontario M1S 3C7; in ENGLAND by T.F.H. Publications, PO Box 15, Waterlooville PO7 6BQ; in AUSTRALIA AND THE SOUTH PACIFIC by T.F.H. (Australia), Pty. Ltd., Box 149, Brookvale 2100 N.S.W., Australia; in NEW ZEALAND by Brooklands Aquarium Ltd., 5 McGiven Drive, New Plymouth, RD1 New Zealand; in the PHILIPPINES by Bio-Research, 5 Lippay Street, San Lorenzo Village, Makati, Rizal; in SOUTH AFRICA by Multipet Pty. Ltd., P.O. Box 35347, Northway, 4065, South Africa. Published by T.F.H. Publications, Inc. Manufactured in the United States of America by T.F.H. Publications, Inc.

Preface

When someone is particularly successful with plants, folklore has it that he or she has a "green thumb." Similarly, one could say that the author of this book has "cold-blooded thumbs." In fact, not only she but her entire family is very successful when caring for cold-blooded animals, be they snakes, geckos, toads, arrow frogs, or fire salamanders. The whole family must be involved, as is obvious from the contents of this book. How else could one explain remarks such as, "crickets, mealworms, and springtails as food animals can be bred in built-in living room cupboards; all that needs to be done is to have slots cut into the back wall and have a light source installed." A statement like this simply presupposes parental participation. Moreover, anyone who insists that mealtimes should be so adjusted that they correspond to the activities of the animals involved does this with the awareness that someone in the family will have to shoulder this task. It is as simple as that.

This book contains many discoveries by the author, and it appears justifiable to say that much research went into it. Regrettably, though, the word "research" is frequently misused. Surrounded by respectability and officially sanctioned competence, an impression arises that without fancy university degrees research is simply not possible. Yet "research" simply means *inquiring*. Inquisitively investigating . . . that is all there is to research, and THAT is what the author is doing. She is curious, enthusiastic, and clever. The questions she raises go beyond the details of terrarium animal care and maintenance; they extend into the animal's environment.

Anyone who has ever spent the family summer vacation in southern Spain so he could study the habitat of the common chameleon notices without fail how environmental factors influence the life of an animal. He also observes the effect man has on the environment, which is rarely ever to the advantage of other living creatures. Environmental and species protection have become as important as daily hygiene. Sadly, it is not a facetious remark when the author suggests not feeding lettuce from a supermarket or green grocer to animals because it contains too much poison.

Please do not misunderstand my remarks; this book is not an elementary textbook for environmental protection. However, the involvement with the "lower vertebrates," which are tied to water and land by their life cycles, touches on our so-called survival instinct. A solid understanding of these animals elucidates important inter-connected relationships. Even if not for their own sakes, the fates of these animals must be viewed as indicators of human excesses upon nature. Therefore, the information conveyed in this book is of interest not just to hobbyists. Moreover, the way this knowledge was acquired may encourage the reader to USE HIS OWN EYES instead of waiting for the observations of others.

Wolfgang Wickler
Seewiesen

The golden harlequin frog, *Atelopus varius*. Photo by K. T. Nemuras.

Contents

Slugs are taken by some of the smaller ground snakes, some turtles, some sala-manders, and a few frogs. When "de-slimed" and chopped into small pieces they are good as a supplemental food for many species of reptiles and amphibians. Photo by C. O. Masters.

Pomacea bridgesi, the common apple snail of the aquarium, provides a large amount of muscle meat for turtles when carefully killed. Its droppings are also a traditional source of infusoria cultures that may be useful for feeding tadpoles. Photo by U. Werner.

Small snails, such as the red ramshorn (*Helisoma*) above, are eaten whole by many aquatic turtles, salamanders, and even by a few frogs. Garden snails, below, will be taken by some skinks and a few other lizards, as well as some snakes. Photo above by U. Werner; that below by A. Norman.

10

Introduction

Amphibians and reptiles, together with fishes, belong to those animals known as the "lower vertebrates." For a long time ignorance and superstition have lead to the belief that particularly the first two groups—amphibians and reptiles— are slimy, disgusting, and dangerous. Consequently, they have not enjoyed the same popularity as other pet animals.

However, anyone who has ever dealt with frogs, newts, lizards, snakes, and tortoises, all cold-blooded (ectothermic) vertebrates, will have soon discovered how interesting their behavior can be. He or she will recognize the complex and diverse reproductive strategies these animals have developed by continually adapting to prevailing ecological conditions in the course of their evolution. These adaptations enable their survival to this day.

For instance, even newts communicate with each other through finely tuned olfactory, tactile, and optical signals during their courtship. This mechanism assures that the female finds and picks up the sperm package (deposited by the male) at the right point in time. Females of some species of poison arrow frogs show such well-developed memory capacity that they are able to find again their own larvae deposited individually into water-filled bromeliad leaf axils throughout the entire developmental period to feed them at regular intervals with specially produced nutritive eggs ("food eggs"). Females of some of the giant snakes (such as *Python molurus*) not only exhibit brood care by coiling around and guarding their clutch, but they are also capable of generating the correct incubating temperature for the eggs.

These are only a few examples of reproductive strategies found in amphibians and reptiles. Their specific requirements, behavior, and breeding are presented in an abridged form in this book. In doing so I am relying in part on the latest published reports covering these subjects. However, apart from that, this book is largely based on results that we—Helmut Zimmermann and I—have obtained during many years of successful care and breeding of reptiles and

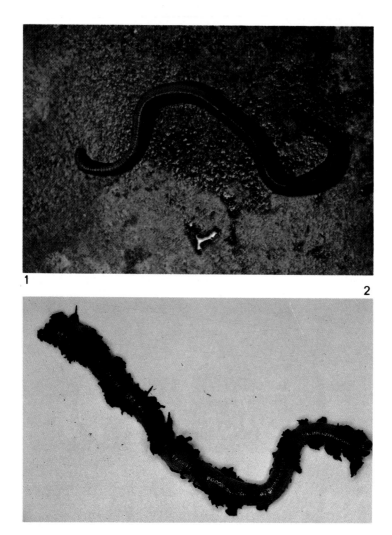

1

2

Common live foods. **1)** Large earthworms, often called nightcrawlers, are acceptable food if they come from soil that has not been treated with insecticides. Photo by C. O. Masters. **2)** Cultivated earthworms are regularly available at smaller sizes all year around and are much safer than wild-collected earthworms as they are sure to have come from chemical-free ground for several generations. Photo by M. Gilroy. **3)** Microworms shown at about double natural size. Photo by M. Gilroy. **4 & 5)** Tubifex (or more properly tubificid) worms are a good food for many amphibians and reptiles and are available in almost all pet shops. They must be kept clean, however; colonies also tend to become established in the substrate of the aquarium. Photo 4 by M. Gilroy; photo 5 by W. Tomey.

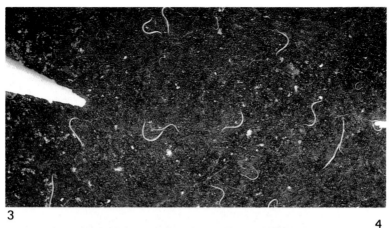

3

4

5

amphibians. Also included in this book are extensive behavioral studies on a large number of terrarium animals. The book also draws on some of our natural habitat studies.

Anyone who applies himself diligently through personal observation and studies of the relevant literature so that he becomes knowledgeable about amphibians and reptiles and their requirements, and who then uses this knowledge correctly, will be successful in breeding these animals, be they miniature versions of glaringly colored poison arrow frogs, the fragile and delicate structures of matchstick-size baby geckos, or entwined masses of beautifully marked young snakes.

This not only satisfies our esthetics and excitement of discovery, but it also helps combat environmental destruction. After all, protection of species and the environment are as important to us as terrarium care. Therefore, it should become a firm goal for all of us to not only keep animals taken from the wild in accordance with their species-specific requirements, but also to breed them regularly and so to conserve these species, at least in captivity. I sincerely hope that this book will be able to persuade you to seek such aims and objectives.

I was encouraged by my publisher to deal in some detail with the topic of breeding terrarium animals. From early childhood on my parents, Helmut and Marianne Zimmermann, gave me the opportunity to gather experience by keeping and breeding animals in captivity, as well as to investigate their ecology and ethology in the wild on numerous trips inland and overseas. My brothers, Peter Zimmermann (botany and conservation) and Axel Zimmermann (measuring technology and electrical engineering), participated extensively on field trips with me. Prof. Dr. H. Rahmann, Zoological Institute, University Stuttgart-Hohenheim, provided invaluable detailed advice. Prof. Dr. W. Frank, Department of Parasitology, University Stuttgart-Hohenheim, reviewed the chapter on diseases. Dr. W. Boehme, Zoological Research Institute and Museum Alexander Koenig, Bonn, reviewed the manuscript critically, in particular the taxonomic section. He also provided valuable suggestions. Prof. Dr. W. Wickler, Max-Planck Institute for Behavioral Physiology, Seewiesen, was willing to support my project with his prefatory comments.

14

The photographers K. Paysan, B. Kahl, H. Zimmermann, and also Prof. Dr. H. Horn, Prof. Dr. W. Sachsse, and Prof. Dr. P. Weygoldt, among others, made available many rare habitat, breeding, and behavioral photographs. I would like to thank all of them for their help.

Last but not least, I would like to express my particular thanks to those authors who, through their published articles, have shared their experiences in keeping and breeding amphibians and reptiles and so made this information available to a large circle of interested individuals.

I am aware that the topic of breeding "terrarium animals" is far from being exhausted by this book. Breeding reptiles and amphibians will gain increasing significance in the future, not only within the spirit of nature protection and species survival, but also as part of a wider public environmental education platform. Consequently, I would indeed be indebted to any reader for constructive criticism, comments, and further suggestions.

Elke Zimmermann
Stuttgart

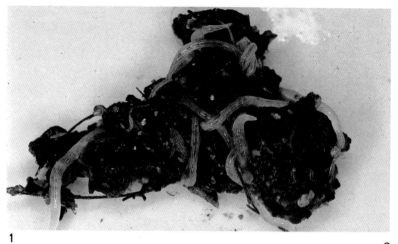

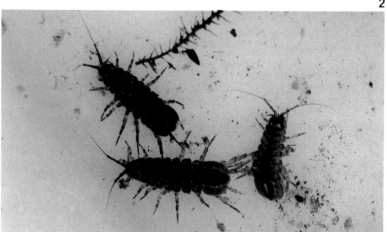

Common live foods. **1)** White worms are readily cultivated at home or purchased in some pet shops and are excellent food for many small herps. Photo by M. Gilroy. **2)** Freshwater isopods or asellids are readily collected in most bodies of fresh water in the Northern Hemisphere and are eaten by many newts and frogs. Photo by M. Gilroy. **3 & 4)** Daphnia (or more properly daphnids) are small crustaceans that are excellent food for larval stages of many salamanders. When collected from the wild (as in photo 3) they are often mixed with possibly dangerous predators and pests, however. Photo 3 by C. O. Masters.

3

4

Housing

There are several ways to keep amphibians and reptiles in captivity. They can be cared for in a suitably arranged and decorated terrarium that can also serve as an elegant room divider, in essence becoming the focal point of the room. Alternatively, a terrarium can be built into a wall and so become an attractive decorative element for the entire room. A technically-inclined hobbyist will incorporate the latest "state of the art" technology into the construction and placement of a terrarium. He will attempt to duplicate the actual environmental conditions and requirements for particular species. Depending upon the species, this could involve the use of preheated and suitably prehumidified air (by controlling the humidity via humidity sensors), an automatic sprinkler system, differential illumination with timer switch and rheostat, and air and water circulating pumps.

Another hobbyist may place greater emphasis on designing his terrarium to resemble a particular habitat by using only those plants, roots, rocks, and sand that occur normally in the area where a certain species lives. Since such "replicas" of nature reduced in scale never quite correspond to the actual proportions of the natural habitat, compromises will have to be made. Often the hobbyist will then, at great expense, take the final step and set up a paludarium (aquarium & terrarium) as large as possible using the latest available technology. Invariably such a display ends up being quite large, usually larger than 2 meters by 3 meters, which of course makes it ineffective for the display of small animals, which would disappear into the background and then be seen rarely, if at all. Further compromises thus will have to be made. The pool area will get a few fish, the swampy area some amphibians and turtles, and the land section gets some toads and lizards. The back and sides are destined to be covered with poison arrow frogs. Hopefully, if all is done correctly, an ecological balance will establish itself.

Anyone who has closely examined even 2 to 20 square meters of natural habitat, whether swamp land, tropical savannahs, or a section of tropical rain forest, for its endemic fauna and flora together with the ecological relationships involved will quickly realize that the paludarium can support only a limited animal population.

What basic prerequisites does a terrarium have to fulfill? Animal keeping in apartments should be controllable; that is, a terrarium must be serviceable easily and without danger of escapes. *The regular occupants and their food animals must not be able to escape.* It is imperative that terrarium animals not only be kept alive but also that they are accommodated according to their species-specific requirements. This then enables them to show a normal behavioral repertoire, which includes their ability to reproduce. Esthetic considerations are largely insignificant to the successful keeping of reptiles and amphibians, and technical refinements are only essential as long as they contribute to the improvement of the environment in captivity.

A prerequisite for providing species-correct conditions in a terrarium is an intimate knowledge of the natural history of an animal and its requirements. This knowledge can be obtained through an extensive review of the relevant literature as well as through detailed habitat studies. Unfortunately this cannot be done for all species kept in terraria. Since the requirements vary from species to species and even from one population of a species to another, exact personal observations of the animal concerned are needed in addition to the facts presented within this book in order to determine the optimal conditions required.

Generally, form, size, and interior decoration of a terrarium or aquarium are dependent upon habitat requirements, the behavior, the size, and the number of animals to be kept and bred. Thus ground-dwelling, burrowing, or digging animals, for instance, need a terrarium with a large floor area and a thick layer of substrate. Climbing and tree- or bush-dwelling species, on the other hand, need a tall terrarium with little floor space but ample climbing opportunities.

Territorial species or those that form a social hierarchy can only be kept in pairs in smaller terraria. If, however, several of these animals are to be kept together, a large terrarium with sufficient hiding places and several feeding sites must be provided. This is the only way these animals can establish

Common live foods. **1 & 2)** Brine shrimp, *Artemia*, are excellent food for any aquatic carnivore. The adults (photo 1) are taken by many newts, while the small nauplii larvae (photo 2) are fine food for small salamander larvae. Photos courtesy Artemia Reference Center. **3)** Glassworms, *Chaoborus* larvae, are just one of several types of fly or midge larvae sold in pet shops. Photo by M. Gilroy. **4)** Bloodworms, larvae of chironomid midges, are available both live and freeze-dried and are excellent food for aquatic animals. Photo by M. Gilroy. **5)** Many types of aquatic insect larvae, such as this stonefly nymph, are excellent food for herps. Just be careful about contamination with pests and predators. Photo by D. Lewis.

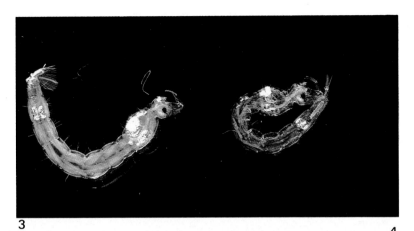

3   4

5

21

territories according to their natural behavior and still avoid contact (i.e., fighting) with each other.

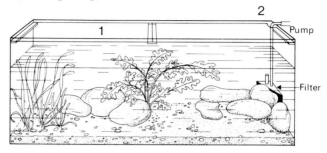

An aquarium set up for aquatic amphibians and amphibian larvae. **1)** Cover with sliding glass panel or a wire mesh frame. **2)** Lighting to be provided by fluorescent tubes.

TYPES OF TERRARIA AND THEIR DECORATION

The following paragraphs provide an overview of the major types of terraria and their setups and decoration. The specific requirements for individual species will be discussed in detail in subsequent chapters. There is already sufficient literature about decorating and planting a terrarium, as well as explicit instructions on how to construct a terrarium, so that these topics need not be dealt with in great detail.

1. Aquaria for aquatic amphibians and amphibian larvae
Substrate: Peat moss layers covered with fine sand or gravel.
Plants: *Riccia, Pistia, Eichhornia, Cryptocoryne, Anubias, Ranunculus aquatilis, Elodea, Myriophyllum,* and others.
Decoration: Hiding places near strategic plant arrangements, such as rock grottos or tangles of vines or aerial roots.
Hardware: For water aeration and cleaning, a circulating pump and filter (however, additional complete water changes are required periodically). Water heater with thermostat for warm-water species.

2. Aquaria with terrestrial sections, including aquaterraria for turtles, aquatic lizards, and snakes, as well as for aquatic frogs and newts
Substrate in water section: Sand or fine gravel.
Substrate in terrestrial section: Sand or leafy soil (coarse compost).
Plants in water section: As listed for *Aquaria* (can be omitted for herbivorous turtles and lizards).

Plants in terrestrial section: *Scindapsus* (ivy arum), *Cyperus* (papyrus), *Philodendron, Phragmites* (rush), and others as available and appropriate.

Decoration: Separate land and water sections with glass panels sealed in place with silicone putty. Access to terrestrial sections with stone bridges (use smooth stones) or tree roots.

Hardware: Water requires aeration. See *Aquaria* for cleaning (filtration) and heating. Infrared radiator for exotic species from warm habitats.

3. Moist terraria for amphibians and reptiles from moist habitats such as forests in temperate and tropical (warm) zones. Relative humidity 70-100%

Substrate: Lowest layer as drainage through medium-sized gravel or similar substances, followed by leafy soil or peat moss, moss pillows.

An aquarium with a land section for aquatic turtles. **1)** Land section in inserted frame made of aluminum or stainless steel. **2)** Substrate for depositing eggs (sand or peat moss and sand mixture). **3)** Spotlight or mercury vapor lamp. **4)** Cover of glass or wire mesh.

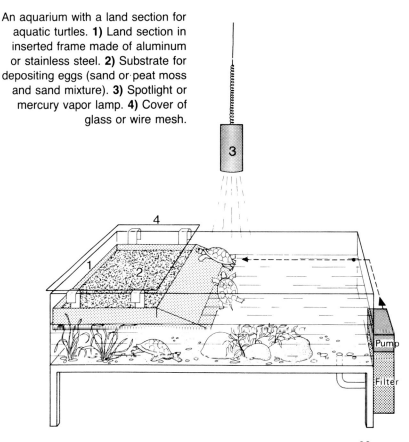

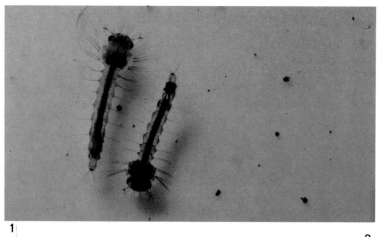

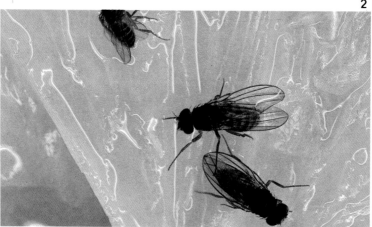

Common live foods. **1)** Mosquito larvae are readily collected during much of the year and make excellent food if the adults are not allowed to emerge. **2)** Flies of all types are excellent food for many reptiles and amphibians. These large specimens (shown twice natural size) would be fine for some adult lizards, salamanders, and frogs. **3 & 4)** The commonly cultured fruitfly, *Drosophila melanogaster.* Both fully winged types (photo 3, shown about double natural size) and vestigial winged mutants (photo 4) are standard foods for many herps. **5)** Maggots of blowflies and similar flies, also called gentles, are suitable food for reptiles and amphibians and are readily cultivated. Photos by M. Gilroy.

3

4

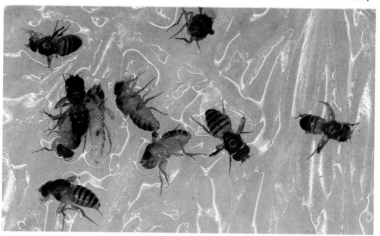

5

A terrarium set up for amphibians and reptiles from tropical rain forests. **1)** Interior back wall made of cork. **2)** Provide fluorescent tubes for lighting. **3)** Cover of glass with wire mesh panels for air flow. **4)** Wire mesh panel for ventilation. **5)** Stream of flowing water from aquarium circulating pump; water should be filtered and the filter cleaned regularly.

Plants in temperate terraria: Ferns, rushes, pewter grass, ivy, tradescantia, sedge, and similar plants.

Plants in tropical terraria: Orchids, bromeliads, aroids such as *Philodendron, Aglaonema, Anthurium, Anubias, Scindapsus,* and *Dieffenbachia,* as well as begonias, ferns, *Ficus,* and similar plants.

Decoration: Back and side walls cork sheeting, styrofoam or peat moss panels, possibly epiphyte-bearing tree trunks or climbing branches. Rocks and pieces of bark or roots as hiding places. Exchangeable water containers; built-in water section by installing with silicone putty a separating glass panel or running water. Cool location along window facing north or east or in basement (temperate moist terrarium) or elevate temperature by means of room heating (tropical moist terrarium).

Hardware: Radiator for lizards, snakes, and tortoises preferring heat.

4. Semi-moist terraria for amphibians and reptiles from semi-moist habitats such as edges of forests, meadows, and similar areas (relative humidity 50 to 70%)

Substrate: Mixture of peat moss and sand or leafy soil (forest soil) with peat moss, possibly covered with peat moss tiles.

Plants for cool habitats: Ferns, rushes, pewter grass, ivy, clematis, albizzia, lilies, wool grass.

Plants for tropical habitats: Citrus, pomegranate, passion flower, *Coffea, Hoya, Ochrosia, Columnea, Chlorophytum, Aphelandra, Asparagus, Oleander, Ficus.*

Decoration: For back and side walls see moist terrarium. Hiding places should be provided by pieces of cork, tree bark, or rocks. Exchangeable water container or installation of dividing glass panel to form distinct water section. Access to terrestrial section through placement of large stones or roots.

Hardware: Radiator required for species from warm habitats, otherwise room heating sufficient.

A semi-moist terrarium. **1)** Stream of flowing water. **2)** Water section. **3)** Circulating pump moving water from water section to "stream." **4)** Glass or wire mesh cover. **5)** Provide fluorescent tubes for light.

5. Dry terraria for savannah-dwelling and desert-dwelling amphibians and reptiles, as well as for species from dry regions in subtropical and temperate zones (relative humidity 50%)

Substrate: Sand; a small damp area should be provided, possibly inside a plant bowl or similar container.

Plants for cool habitats: *Sedum, Thymus,* rushes, club rushes, heath.

Plants for warm habitats: Agaves, acacias, *Genista,* echeverias, yuccas, cacti, *Euphorbia, Saxifraga, Podocarpus,* grasses.

1

2

Common live foods. **1 & 2)** Adults and larval mealworms, one of the "old stand-bys" for feeding lizards and other herps. Many keepers advise caution when feeding mealworms and suggest never feeding *only* mealworms. **3)** A waxmoth larva of the type occasionally seen for sale. **4 & 5)** Crickets, a familiar and reliable food. The exact type available will vary from season to season and supplier to supplier. Photo by M. Gilroy.

3

4

5

Decoration: Roots, branches, cacti skeletons, or stones and rocks as hiding places; back wall possibly of cork. Small water bowl.

Hardware: Radiators and floor heating for reptiles; UV-A tubes for reptiles.

THE OUTDOOR TERRARIUM

An outdoor terrarium provides the most ideal maintenance and breeding conditions for northern temperate reptiles. Even species from the subtropics as well as desert and savannah inhabitants can easily be kept in such a setup during the summer months. There are a number of different types of outdoor terraria just as there are different types of indoor terraria. These are described here briefly.

Outdoor terraria for lizards, snakes, and amphibians can be made of bricks or concrete. They must be at least 1 meter high and be protected against the intrusion of cats, rats, and birds. The animals in such terraria must be prevented from escaping by placing a wire mesh top over the enclosure.

The lowest substrate layer should consist of gravel or expanded clay in order to provide drainage. This is followed by a layer of top soil and leaves or sand. There must also be ample hiding places in the form of stone or rock structures, climbing branches or roots, and even tree trunks. The plants used depend upon habitat and specific requirements of the animals.

There are even simpler structures for turtles and tortoises. A large and sufficiently deep plastic wading pool or similar container that can be equipped with a filter is placed into the ground to provide suitable outdoor accommodation for many turtles. A large piece of root or similar object serves as access to the land section. This rock is also frequently used by the turtles to bask in the sun. A wire mesh cover with its corners buried in the ground prevents the turtles from escaping and at the same time provides protection against cats and birds. Tortoises from subtropical regions can be kept in most backyards by simply housing them in an area enclosed by a wooden fence and covered with wire mesh.

All outdoor terraria must have adequate protection against the elements, especially against the sun and rain. Depending upon the species, there may also be requirements for moist or swampy areas as well as for dry areas. If the soil layer beneath the terrarium is sufficiently deep so that the animals can actually burrow and thus find frost-free areas, they can even be kept outdoors

A dry terrarium for amphibians and reptiles from arid regions. **1)** Plant section with substrate kept sligtly damp for depositing eggs. **2)** Water bowl. **3)** Lighting by spotlight or mercury vapor lamp, or possibly by UV-A tubes. **4)** Wire mesh cover.

A terrarium for tortoises. **1)** Plant section with substrate kept slightly damp for depositing eggs. **2)** Open box enclosure with dry substrate for sleeping and hiding. **3)** Lighting with spotlight or mercury vapor lamp.

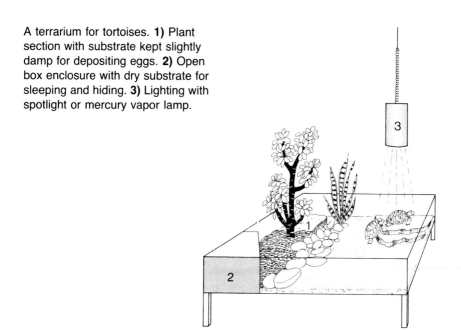

Perhaps the most common food for pet snakes and some larger lizards is the domestic mouse. Recently born young (above) are often called "pinkies." Frozen mice are often available at the pet shop and are much more convenient than trying to culture mice yourself, unless you have enough room and the spare time. Photo above by E. Jukes; that below by Dr. Herbert R. Axelrod.

right through mild winters. However, it is less risky if the animals are taken back indoors and kept in "hibernating boxes" or indoor terraria. These can be placed in a cool basement or storage room and maintained at an air temperature of from 5 to 10 °C, but the night temperature must *not* drop below 5 °C. The actual length of hibernation is species-specific, and among those species with a wide distribution it even varies from one location to the next. It can range from two to six months.

The most suitable substrate for hibernating tortoises (as well as for many lizards, snakes, salamanders, newts, frogs, and toads) is a mixture of leaves and moss. For sand-dwelling animals such as spadefoot toads *(Pelobates, Scaphiopus,* etc.) smooth river sand is good. It is important that the over-wintering cage is sufficiently ventilated, yet the animals have to be protected with a wire mesh lid against rats and mice. The substrate should always be kept slightly moist or damp.

TERRARIUM CONSTRUCTION

Some of the most frequently used materials for building a terrarium are glass, aluminum, fiberglass, and wire mesh. Wood is less suitable due to its low heat and humidity resistance. Glass terraria can be built with little effort and at low cost. Moreover, they are convenient to monitor and easily cleaned and disinfected. Therefore, the construction of a basic terrarium is described here briefly. Remember that length measurements have to take into account the lengths of the fluorescent tubes to be used. Many standard all-glass aquaria that are available cheaply at petshops will serve admirably as terraria.

Five sheets of glass (bottom and four sides) about 5 to 10 mm thick are pre-cut to the required size and the edges smoothed off with very fine sandpaper. Then all edges and adjacent areas are carefully cleaned off with lighter fluid, acetone, or a similar fat-dissolving fluid. After that, the four sides are set up vertically and supported by bricks, sticky tape, or any other device that will hold the glass temporarily in position. In order to achieve smooth and even seals, match sticks are placed between the sheets of glass to keep them equally apart along their entire joints. The joints are then filled with silicone putty, doing the four sides first and then the bottom glass, which is placed on top of the four sides (here too the match stick spacing should be maintained). About one day later, after the sealant has hardened, the terrarium is turned over and placed on its bottom glass panel.

Now additional installations can be made, such as divider panels, U-tracks for sliding tops, and others. The match sticks are now removed (pulled out) and the holes thus created are filled with fresh silicone. This standard module can of course be changed in any number of ways in order to fit a particular situation (species, habitat, or location).

LIGHT AND HEAT

Amphibians and reptiles are poikilothermic; that is, they have to be warmed up to their optimum body temperature by an external heat source in order to be able to feed, digest, and reproduce.

Unfiltered sunlight used as illumination as well as a heat source provides the best possible conditions for keeping and breeding those reptiles that prefer direct solar radiation (amphibians tend to avoid prolonged exposure to direct sunlight.) However, not everybody can keep his animals in outdoor terraria during the summer months. Even placing the indoor terrarium so that it gets at least a few hours of unfiltered sunlight a day is often not possible. Therefore, many hobbyists have to rely on artificial illumination and heating.

Fluorescent tubes as an artificial light source are closest to the color spectrum of sunlight. They give off sufficient light but little heat at relatively low energy consumption. Not all fluorescent tubes are suitable for maintaining reptiles, though many brands are. I have found that tubes such as the Philips TL-D Super 80, Osram Lumilux, Sylvania Grolux, Osram-L-Fluora, True-Lite, and Verilux-True Bloom provide the most suitable combinations for illuminating reptile and amphibian terraria. We have also found the Philips TL 40W/09 and Osram-L 40-100W/79 to be very useful for those lizards that like direct sunlight. For instance, some toad-headed lizards—which the literature describes as being difficult to keep—are still maintaining their complete cycle of activities after four years in our terrarium with continuous illumination from 08:00 hrs. to 20:00 hrs. with these types of lamps. Similarly, our *Callisaurus draconoides* are also still highly active after two years. Israeli fringe-fingers *(Acanthodactylus boskianus),* also desert-dwellers, have already bred successfully several times. Damage due to exposure to the UV-A tubes could not be observed in any of these animals or in savannah-inhabiting marble frogs that were exposed to such long-term radiation.

If UV-A tubes are not being used, one has to provide regular supplementary radiation with fluorescents equivalent to brands

34

such as the Osram Ultra-Vitalux, at least during the late fall, winter, and spring months (see under rickets in the disease chapter). Only the combined effects of UV lights, vitamin D, and calcium supplements will assure healthy bone structure in these animals. However, radiation in this spectral range is damaging to plants, so they must be removed.

Fluorescent tubes can be mounted inside a hood lined with polished aluminum for better reflection and placed over the terrarium. Various types of hoods are available at petshops or can be constructed. If the heat given off by the ballast is to be utilized as a supplementary heat source, the ballast should be mounted underneath the terrarium.

Mercury vapor lamps that are operated through a transformer provide a great deal of light but also give off a large amount of heat, so their use is really only economically viable for very large terraria. Incandescent light bulbs with built-in reflectors and also spotlights give off a lot of heat and light and can be utilized to provide localized heating for specific areas, such as a basking rock.

Simple heat sources are heating coils with plastic or lead insulation or heating elements that can be thermostatically controlled. These heat sources are used primarily in dry terraria for savannah- and desert-dwelling reptiles that require additional radiating heat from below. Infrared radiators are used only in supportive treatment of sick reptiles. These units provide potent, penetrating heat.

Aquaria and aqua-terraria are heated by means of thermostatically controlled aquarium (glass-tube) heaters. These same heaters can also be used to warm up the water section in a terrarium if the depth of water is great enough to allow the sensing element of the thermostat to be covered.

All light and heat sources have to be installed in such a way that the terrarium inhabitants cannot come into direct contact with them; this then avoids burn injuries. Timer clocks can be used to control light-dark changes as well as temperature fluctuations between day and night (when natural daylight is not available). A dimmer switch or rheostat can provide for a short period of dusk to give the animals an opportunity to find their sleeping places before total darkness sets in.

Daily or seasonal variations of temperature, relative humidity, light intensity, water depth, or length of illumination are particularly important for many amphibians and reptiles. These factors often control the onset of breeding, duration of the breeding cycle, periods of activity, rest periods, and other physiological and behavioral functions.

HUMIDITY AND WATER

According to their natural habitat, terrarium animals have variable requirements for humidity. Those species inhabiting perpetually damp habitats such as rain forests need a high humidity, from 70 to 100%. Therefore, the terrarium has to be sprayed several times a day with a plant mister or a small self-priming high pressure pump (because of the danger of calcium carbonate deposits inside the pump, only water passed through an ion exchanger should be used). Those hobbyists who are mechanically inclined possibly can construct an automatic misting system with a hygrostat and timer switch. The simplest method for creating a constantly humid environment in a terrarium is to restrict ventilation to the barest minimum. This would permit breeding some of the dendrobatid species (poison arrow frogs) in aquaria where the wire mesh top has been closed or replaced by a glass cover, leaving only a narrow gap open for ventilation. However, the stagnant air inside the tank provides for a microclimate that most plants do not tolerate very well.

The humidity inside a terrarium can also be increased by placing a thermostatically controlled aquarium heater in the water section of the terrarium or by creating a constant circulating water flow (waterfall, stream) with a small circulating pump. Large terraria are best served with a commercially available humidifier.

VENTILATION

Air circulation into and out of a terrarium is best accomplished and controlled by adjustable ventilation areas of gauze or wire mesh on the top or side walls of the enclosure. Fluorescent tubes tend to heat the air inside the container, which then rises and escapes through the ventilation areas at the back of the top cover. At the same time fresh, cooler air enters via the ventilation areas along the sides and flows along the bottom. If there are no lateral ventilation openings, an aquarium aerating pump can be used to supply fresh air to the lower portion of the terrarium. Moreover, this way the air can be warmed up and humidified.

Automatic water changers are now readily available and are producing very good results when used with fishes. Since salamanders and frogs—not to mention turtles—are notoriously "dirty" aquarium inhabitants, automatic water changers should prove to be especially valuable in the aqua-terrarium.

Nutrition

With the exception of a few highly specialized species, a widely diversified food spectrum is an absolute prerequisite for optimum maintenance and breeding of terrarium animals. Although we are rarely ever in a position to offer our animals the food found in their natural environment, we can, however, make substantial efforts to collect, catch, buy or breed the various food animals and acquire the needed food plants, thus assuring the required variable diet. It is precisely such a diversified food supply together with feeding live food animals to insectivorous, carnivorous, and omnivorous terrarium animals that stimulates them to proceed with their normal activities in the terrarium and so catch their own prey.

Only sick and weak animals are fed by means of food wires, needles, or forceps, and then only by using pieces of food enriched with multivitamins, calcium preparations, and—if need be—medication.

If food is given in small bowls or other containers, one has to watch that each animal gets sufficient food. Frequently weaker or lower ranking animals are pushed aside by stronger or higher ranking animals. It may be advisable to set up several feeding places; these should provide feeding opportunities to weaker or smaller animals without visual contact with other cage inhabitants.

Feeding times always depend upon activity periods of the species concerned. The amount of food required depends upon a number of factors, such as size, ambient temperature, movement and activity, and the developmental and nutritional condition of an animal. Pregnant females and young animals always have higher food, calcium, and vitamin requirements.

Vitamin supplements are given to reptiles sprinkled over their food, mixed in with the food, together with food organisms, or through injection of liquid vitamin solutions into food insects or freshly killed mice about to be given. The re-

38

quired vitamin preparations are available from petstores or medical practitioners. There are quite a few preparations that can be used for the treatment of and prevention of nutritional deficiencies in amphibians and reptiles (though exact scientific research on their effectiveness is usually available only from results obtained from human or pet nutrition).

Water-soluble multivitamin preparations as well as calcium lactate, a water-soluble calcium preparation, can be given to reptiles in their drinking water and to amphibians in their bath water. Many vitamin and mineral preparations in powder form, as well as vitamin D3 preparations suitable for the treatment of acute rachitis and supplements of specific vitamins, are usually fat-soluble and can be given sprinkled over the food or injected into it.

NATURAL FOODS

Herbivorous and omnivorous reptiles (those feeding on plants and those that take in a large portion of plant material but will also eat animal protein) can be fed on a large variety of wild plants, provided that they have not been sprayed with insecticides or weed killers. The list of suitable plants is long and includes dandelions, clover, coltsfoot, dead nettle, chickweed, and others. Also often taken eagerly are buds, flowers, and fruit of non-poisonous wild flowers and plants, bushes, shrubs, or trees from the backyard.

Meadow plankton (a collective term for all insects caught in a fine-meshed insect net) is diverse and rich in vitamins. It is taken eagerly by most small to medium terrarium animals. However, those food items that are either too small or too large and thus remain uneaten will quickly contaminate the terrarium. Moreover, with natural live food there is also an inherent danger that diseases may be carried into the terrarium.

Salamanders, newts, turtles, and some of the smaller snakes can be fed on earthworms dug from the top soil of lawns, meadows, or fields (NOT from compost or dung heaps!). These worms represent valuable calcium-rich food. Small slugs that inhabit shady and damp areas in any backyard are eagerly eaten by newts, salamanders, some frogs, toads, and small snakes. Snails occurring in dry regions are taken by skinks and certain other lizards.

Mosquito larvae and a variety of aquatic crustaceans can be

collected from spring to fall in clean pools, ponds, and lakes by using a fine-meshed aquarium net or plankton net. These organisms serve as food for aquatic amphibians, some amphibian larvae, and some turtles.

Leaf lice (aphids) picked off fresh plant shoots and leaves that have not been treated with an insecticide, springtails, and small spiders are suitable as rearing food for newly metamorphosed terrestrial frogs and salamanders and also for newly hatched lizards. Flies of all kinds can be attracted from spring to fall by using raw meat or fish as bait. Commercially available fly traps (not those using chemical attractant or other chemicals) have also proved to be very effective. Nocturnal moths are eagerly taken, especially by geckos. These insects are easily attracted at night to a light source.

For the sake of environmental protection, those reptile species that feed exclusively on amphibians or other reptiles should not be maintained in captivity and are not discussed in this book except in those cases where food animals can be bred in sufficient numbers.

COMMERCIALLY AVAILABLE FOODS
Petshops, biological supply houses, bait shops, and similar establishments can provide a wide range of live and dead food animals for terrarium animals. Most commonly available are mice, rats, and other small mammals, fish, newly hatched chicks, crickets, mealworms, tubifex worms, waterfleas (daphnia), and cyclops, as well as various dried foods for raising tadpoles or for feeding turtles. Fly maggots (gentles) can sometimes be purchased in bait shops. After having been purchased in a shop, flies, mealworms, and crickets should be given food for another day or so before they are offered to terrarium animals. Supplementary foods for carnivores and omnivores are lean beef with excess fat trimmed off, heart, and liver, as well as fat-free minced meat. To the latter one can conveniently add multivitamin and mineral preparations. Eggs are keenly taken by egg-eating snakes and some lizards. We found egg yolk (because of its high nutritional value) very useful for raising the tadpoles of some tropical frogs, such as *Dendrobates histrionicus, D. lehmanni, D. pumilio,* and *D. speciosus.*

Different vegetables and fruits can also be purchased or (preferably) raised and harvested in your own backyard.

BREEDING FOOD ANIMALS

If our terrarium animals are to receive a diversified diet throughout the year and we wish to be independent from any market supply of food animals, we have to raise the foods ourselves. There are different ways of doing this, therefore the following has been confined to the most common food sources and the respective breeding methods and procedures that have become well established and proven.

Cleanliness is of paramount importance not only for the care and maintenance of terrarium animals, but also in the production of food animals.

The breeding setups mentioned below (with the exception of enchytraea, whiteworms) should be accommodated in a dry, temperate room. Boiler rooms, greenhouses, or something similar would be most suitable. However, a built-in closet in an apartment can also be converted to accommodate food animal breeding facilities (not acceptable for mice!!!). For that purpose the sides of the closet are insulated with styrofoam sheets or tiles about 2 to 3 cm thick. Commercially available sheets are cut to the correct size and glued together with special styrofoam glue. The setup is illuminated with a fluorescent tube mounted on the ceiling of the closet or with an incandescent bulb inside the breeding container. In order to generate some heat, an incandescent light bulb also should be mounted at the top or bottom of the closet. Slots along the top and bottom of the back wall, when covered with wire mesh, provide ventilation.

An insulated heated cupboard for breeding food animals (after Zimmermann, 1982). The depth is about 45 cm.

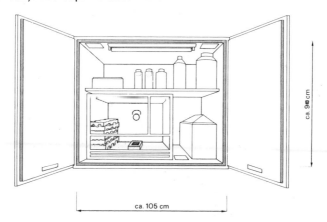

ca. 90 cm

ca. 105 cm

Details of breeding boxes suitable for crickets and flies, after Zimmermann, 1982.

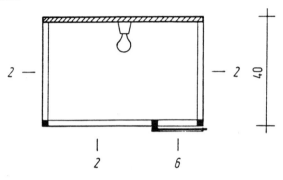

A.

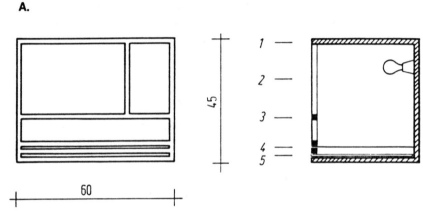

A. Schematic views of the construction of the basic box in front view (bottom left), cross section (bottom right), and from above (top). 1) Masonite panel. 2) Wire gauze. 3) Hardwood. 4) Wire mesh. 5) Metal tray. 6) Sliding Plexiglas lid.

B. Outside view of the finished insect breeding box (left) and a rearing container (right) suitable for grasshoppers and crickets.

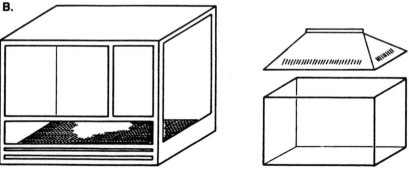

Breeding grasshoppers, crickets, and flies is most productive in containers constructed and set up as follows: The back and bottom of a frame structure (60 cm x 40 cm x 45 cm, size dependent upon space availability) are covered with Masonite sheets glued to each other; the two sides and the top are covered with metal gauze or fine wire mesh. The front consists of the same mesh material except for a small opening for a Plexiglas or acrylic sliding panel. This panel should slide in grooves but must give a tight seal against the mesh. The sliding cover provides access to the breeding cage; all food animals and materials (food, water, etc.) have to come through this opening, which has to be sufficiently large. Cleaning is facilitated by installing a wide-mesh (5 x 5 sq. mm) wire bottom with a removable tray underneath.

If newly hatched grasshoppers are to be raised in such a container, the wire mesh bottom is replaced with an equally thick plastic bottom, so that the larvae do not fall through to the bottom tray.

Feeding of grasshoppers and crickets should follow the breeding table. However, if lettuce from the local supermarket is used as food, every precaution has to be taken that it has not been treated with insecticides. Once we unknowingly fed contaminated lettuce that overnight wiped out our entire grasshopper and cricket broods contained in about 20 boxes! Since then we have changed from feeding lettuce to germinated wheat (which we produce ourselves) during the winter months. A well-soaked, thick layer of newspaper is covered with a layer of wheat grain, which is permitted to germinate to a height of about 10 cm. This setup should be kept in a well-lit location and maintained warm and moist (through regular sprayings with a misting bottle). Germinated wheat is not only a highly nutritious protein-rich and vitamin-rich food for grasshoppers and crickets, but it is equally eagerly eaten by omnivorous and herbivorous lizards and turtles.

In conclusion, there is one important point that has to be kept in mind when breeding food animals for reptiles and amphibians: special care must be taken that the food animals *do not escape*. This may save a lot of aggravation in terms of possible disputes with neighbors, complaints and damage claims from landlords or neighbors, etc.

KEEPING AND BREEDING IMPORTANT FOOD ANIMALS

White worms *(Enchytraeus albidus)*
Container: Wooden box (about 20 10 10 cm) with glass cover.
Maintenance: Moist peat moss/soil mixture. AT★: 10 to 15° C; RH★: 70%; dark location.
Food: Every third day about one teaspoon firm porridge (oatmeal) buried in the soil.
Availability: Can be purchased in petshops, biological supply houses, advertisements in aquarium journals.
Remarks: Breeding container must never be permitted to dry out. Leftover food to be removed every third day.

Fruitflies *(Drosophila melanogaster* **and mutants, other available species)**
Container: 250-1000 ml beakers or jars covered with fine gauze (nylon stockings or similar material).
Maintenance: Cover bottom of container with about 2-cm-thick layer of food medium; provide climbing facilities by placing sticks of filter paper, wood shavings, tongue depressors, or similar material on top of medium. AT: 25° C; RH: 80 to 90%.
Food: Food medium made of milk and farina or similar cereal, mixed with a few grains of dry yeast and some liquid multivitamin. Sprinkle some fungus inhibitor over the mixture in order to prevent fungal growth. For further details consult specialized manuals for recipes for fruitfly media.
Availability: Commercially available from biological supply houses (who will often supply culturing instructions). Often sold through ads in aquarium magazines. Can be attracted by decaying fruit during summer months.
Remarks: Breeding setups must be replaced with every new generation. Development period from egg to adult is 15 to 21 days.

Crickets *(Gryllus bimaculatus, Acheta domestica)*
Container: Plastic terrarium with lid; commercially available.

★AT = Air Temperature; RH = Relative humidity

44

Maintenance: Newspaper, rolled-up paper, or empty egg cartons piled on top of each other to provide hiding places; require normal daylight or artificial illumination. AT: 30° C; RH: 30 to 40%.

Food: Petri dishes or similar containers with fruit such as apples or pears; vegetables such as finely cut carrots, cabbage, or cucumbers; crushed oats (uncooked oatmeal) or bread; lettuce; germinated wheat; grass, dandelion, or clover.

Availability: Commercially available through bait shops, biological supply houses, advertisements in aquarium and terrarium journals.

Remarks: Food and food containers must be refilled and replaced daily. To breed adults, fill a flowerpot or other container about 8 cm high, diam. 10 cm with moist sand/peat moss mixture and place in rearing container. Development period from egg to nymph is about 1 week; nymph to adult is about 7 weeks. Nymphs 5 days old can be separated into different size groups if desired. *Acheta* has a 2-week cycle from egg to nymph, 8 weeks from nymph to adult; transfer from breeding container in 10 days.

Laboratory Mice *(Mus musculus)*

Container: Commercially available mouse breeding cage with wire cover (several models available).

Maintenance: Sawdust, wood shavings, vermiculite, etc. Place soft paper tissue in cage when young are born to provide nesting material. AT: 25-28° C; RH: 30-40% (some recommend AT: 22-25° C; RH: 50-70%).

Food: Pellets for mice and rats are commercially available from many petshops. Will also eat fruit, bread, nuts. Uneaten food must be removed promptly. Water should be given in special rodent drinking bottles.

Availability: Available from petshops or through advertisements in relevant journals.

Remarks: Cage should be completely cleaned out and bedding replaced 1 or 2 times per week. Gestation period 19 to 21 days. Litter size 6-12. Weaned after 21 days; young should be removed at that time. Sexual maturity in about 6 weeks. Breeding pair should be replaced in about 10 to 12 months.

Mealworm *(Tenebrio molitor)*

Container: Plastic container about 42 × 26 × 16 cm or similar size covered with fine wire gauze.

Maintenance: Fill container halfway with mixture of crushed oats and wheat, covered with newspaper. AT: 26° C; RH: 30%. Normal morning and evening (diurnal) lighting (daylight or artificial) required.

Food: Small amounts of carrots, apples, and other fruit and vegetables cut into slices. Yeast, bread, and dried meat also taken. Food must be relatively dry and in small pieces. Beware of fungus.

Availability: Available from petshops and biological supply houses as well as advertisements in relevant journals.

Remarks: Fresh food must be replaced every 2 to 3 days. Development: egg to larva about 6 days; larva to adult 4 to 7 months.

Springtails (Collembola)

Container: Glass or plastic container with cover.

Maintenance: Moist sand/peat moss mixture as substrate (up to about ⅓ height of container), dried leaves, tree bark. AT: about 22° C; RH: about 100%.

Food: Small amounts of yeast, dried leaves, tree bark, moss, decaying plant material, slices of carrot, apple, potato, etc.

Availability: Collect specimens under flowerpots, among decaying plant matter, under boards and heavy branches lying on the ground, under bark or moss.

Remarks: Substrate must always remain moist to damp.

Housefly (*Musca domestica*)

Container: Wooden (Masonite) boxes about 60 x 40 x 45 cm. Sides and lid made of fine wire mesh. For small setups, gauze-covered glass or plastic containers can also be used.

Maintenance: Hard plastic refrigerator containers (about 10 × 10 × 8 cm) or similar containers can be used for depositing eggs if filled ⅔ with food medium. AT: 27-28° C; RH: 60%. Diurnal illumination.

Food: Food media for larvae: crushed oats/wheat, uncooked oatmeal, milk powder, cottage cheese with a few grains of dried yeast thoroughly mixed with water (well-kneaded). Food for adult flies: small dish with unsweetened condensed milk and minced meat mixed with multivitamins.

Availability: Obtain from biological supply houses and some bait shops. Attracted through placement of old meat, etc., outdoors in open places.

Remarks: Food medium must always be kept moist. Development time: egg to adult about 18 days.

Waxmoth *(Galleria mellonella)*

Container: Hard plastic container about 20 × 20 × 10 cm equipped with tight-fitting lid and gauze insert.

Maintenance: AT: 25-28° C; RH: 30%. Replace honeycombs or substitute mixture as required.

Food: Honeycombs from beekeepers or substitute mixture of 125 g honey, 125 g glycerin, 125 g ground wheat germ, 75 g yeast (flaky), 500 g poultry meal, 500 g cornmeal.

Availability: Apiarists (beekeepers) or advertisements in fishing magazines.

Remarks: Development time: egg to larva about 14 days; larva to adult about 42 days.

Grasshopper *(Locusta migratoria* **and similar species)**

Container: Breeding box (60 × 45 × 40 cm) with light bulb for heat and raised wire mesh floor.

Maintenance: Branches for climbing. Fill container for breeding (10 × 20 × 8 cm) with mixture of damp peat moss and sand. AT: 35° C; RH: 30%.

Food: Petri dish or similar container with bran, grass, lettuce, or germinated wheat.

Availability: Commercially available through advertisements in hobby journals; also some biological supply houses.

Remarks: Substrate in breeding container must be kept moist at all times. If separation into different sized groups is required, transfer content to new container after about 2 weeks. Development time: egg to nymph about 2 to 3 weeks; nymph to adult about 2 months.

Diseases

Newly acquired animals represent a serious health threat to an established collection. Therefore, right from the start we must make it a mandatory requirement that every new animal—as well as any sick animal from within our established collection—is placed into a quarantine terrarium immediately for better control and observation. A quarantine period should last from two to three months. The quarantine terrarium must be largely sterile and be easily cleaned. For lizards, snakes, and tortoises, the bottom of the quarantine cage can be covered with several layers of newspaper; amphibians should be given a foam rubber substrate. Food intake, defecation, shedding of skin, and behavior of each animal should be closely observed. In reptiles all soft body areas, the skin beneath the scales, the extremities, and around the eyes and ears should be checked for mites and ticks. If infested by these arachnids, treatment must be initiated promptly following instructions given below. Fecal samples from amphibians and reptiles should be taken regularly to a veterinarian, parasitologist, or bacteriologist for examination. Fecal samples and subsequent bacteriological cultures can reveal parasites and disease-causing bacteria of the digestive tract so a specific treatment can be initiated.

Hygiene is of paramount importance for the quarantine cage as well as in all terraria. Therefore, particular cleaning equipment is used in one specific terrarium only. Moreover, the cleaning equipment must be thoroughly cleaned and possibly sterilized after each usage. Dirt and debris such as feces, shed skins, and similar material should be removed promptly. The drinking and bathing water should be replaced daily if possible. The terrarium should be cleaned thoroughly. Food animals must never be transferred from one cage to another because they can be potential disease vec-

48

tors. After working in or on a terrarium it is important to wash your hands immediately (possibly with a disinfectant).

For weakened and exhausted animals that refuse food, force-feeding is the only viable alternative in order to keep them alive. To do this, the mouth is opened gently and a suitable dead food animal or a piece of meat (possibly pre-treated with multivitamin preparations or medication) is inserted. When force-feeding snakes it is advisable to dip the food item into raw egg white so that it slides more easily down the digestive tract. Immediately after the food has been inserted it should be massaged (gently) into the stomach so that it cannot be regurgitated again right away.

The most common diseases of reptiles and amphibians are listed in the table. This table is based primarily upon examinations and histological sections made at the Institute. Also taken into consideration were lectures by and discussions with the head of the Department of Parasitology, University of Hohenheim, Prof. Dr. Werner Frank. I would like to take the opportunity to express my gratitude to him at this point. Suggested disease treatments indicated with an asterisk (*) should be undertaken only by a veterinarian or a parasitologist.

SUMMARY OF COMMON DISEASES IN AMPHIBIANS (A) AND REPTILES (R)

I. BACTERIAL DISEASES

Abscesses (A & R)
Cause: Injuries with subsequent bacterial infections.
Symptoms: Skin lesions or papules with yellowish liquid or solid contents (can be surrounded by connective tissue capsule).
Treatment: (*) Surgical removal with antibiotic application; sulfonamides for amphibians.

Eye infection (R)
Cause: Presumably bacterial infections caused by keeping the animal too cold or in drafty conditions, etc.; in turtles also avitaminosis (see vitamin A deficiency).
Symptoms: Infected, swollen eyes; eyes "glued" shut.

Treatment: (*) Heat treatment, antibiotic eye ointment combined with multivitamin supplements.

Lung infection (R)
Cause: Bacterial infection (*Pseudomonas, Pneumococcus,* and others) often connected with incorrect maintenance (drafts, kept too cold, etc.).
Symptoms: Difficulty breathing; foam exuding from mouth and nose.
Treatment: (*) Infrared radiation (animal must be able to move away from heat source), injections with antibiotics or sulfonamides, or oral antibiotics such as doxycycline on 6 successive days.

Stomach/digestive tract infection (R)
Cause: Bacterial infections (*Pseudomonas, Aeromonas,* and others).
Symptoms: Refusal to feed; soft, foul-smelling feces in snakes; often regurgitation of partially digested food.
Treatment: (*) Chloramphenicol 50 mg/kg BW (Body Weight) for 10 days, reduced to 30 mg from the third day on. Sulfamethazine: 6.5 g/L for 10 days in drinking water together with oral administration of aureomycin (220 mg/kg BW).

Mouth rot (R)
Cause: Bacterial infection (*Pseudomonas, Aeromonas,* and others) combined with general weakening condition; vitamins A and C deficiency.
Symptoms: Purulent inflammations of the mucal membranes; purulent areas and lesions on the jaws.
Treatment: (*) Affected skin areas are painted with sulfamerazine or sulfameter; removal of detached purulent material and skin layers with cotton wool or injections with oxytetracycline (25-50 mg/kg BW) on three successive days.

External fungal diseases (A & R)
Cause: Various *Mucor* species, *Rhizopus, Cephalosporium,* and others.
Symptoms: Changes in the appearance of the skin. For example, in snakes brown skin patches appear that later begin to hemorrhage.

Treatment: (*) Difficult. Antimycotic preparations. Amphibians: Painting with 8-hydroxyquinoline (1:5000) or equivalent. Reptiles: Dabbing with diamethazole. Turtles: Potassium permanganate baths (1:100,000).

Red-leg disease (A)
Cause: Bacterial infection *(Pseudomonas hydrophila).*
Symptoms: Reddening of skin, hematomas along abdominal region and on legs.
Treatment: (*) Sulfamethazine bath (30 ml/20 L water) for several weeks or sulfanilamide bath (200 mg/L) for about 3 to 4 weeks.

Salmonellosis (R)
Cause: Bacterial infections of digestive tract (various *Salmonella* species).
Symptoms: Possibly diarrhea (intestinal infection); however, rarely are specific symptoms present. Approximately 40 to 60% of all reptiles are infected and frequently latent carriers. It is imperative that proper hygiene is observed, since these disease organisms can possibly be transferred to humans.
Treatment: (*) Complete freedom from *Salmonella* bacteria is rarely accomplished. Chloramphenicol (50 mg/kg BW) can be tried.

Dropsy or tympanitis (A)
Cause: Metabolic malfunctions; presumably also bacterial infection.
Symptoms: Tissue fluid accumulations under the skin; bloating of abdominal cavity.
Treatment: Difficult. Sometimes puncturing the tympanum with a disinfected needle alleviates the problem. Continuous fresh water inflow, gentle massages, or forced baths in nitrofurazone for several hours.

II. PARASITIC DISEASES

Amoebiasis (A & R)
Cause: Entamoeba invadens (R) or *E. ranarum* (A).
Symptoms: Detectable through fecal samples followed by cul-

tures. External symptoms are non-specific: excessive fluid intake, bloodied diarrhea, constipation, abnormally extended posture. Internally there is tissue destruction.

Treatment: (★) Clont® (3-5 times human dosage) on 7 successive days.

Coccidiosis (R)
Cause: Eimeria, Isospor, and other coccidian protozoans in digestive tract, gall bladder duct, and bladder.
Symptoms: None externally.
Treatment: (★) Sulfamethazine (6.5 g/L water) for 10 days in drinking water.

Dinoflagellate and ciliate infestations (A)
Cause: Oodinium, Trichodina, and others.
Symptoms: Gray epithelial layer on gills and skin of aquatic amphibians.
Treatment: Copper sulfate (2 mg/L water); trypoflavine (10 mg/L water).

Flagellate infestation (R)
Cause: Leptomonas and others.
Symptoms: Colitis; in chameleons often not pathogenic.
Treatment: (★) Same as for amoebas.

Mite and tick infestations (R)
Cause: Blood mites (such as *Ophionyssus, Liponyssus,* and others); ticks from the families Ixodidae and Argasidae.
Symptoms: Rubbing against objects in cage, scratching, continuous bathing, loss of blood (anemia). Blood sucking activities of mites may occur only at night. Mites may be inactive during the day, hiding in cracks, crevices, and other cage objects.
Treatment: (★) Thorough disinfecting of entire cage, as with trichlorfon. (Caution! Highly poisonous!)
—*Mites:* Place affected animal in small cotton bag that has previously been sprayed with .2% trichlorfon solution; or suspend plastic insecticide strip (pest-strip) in terrarium; or apply Odylen to skin of animal.
—*Ticks:* Kill through suffocation in ether, oils, or petroleum jelly, followed by manual removal (through partial rotation when pulling organism out); subsequent treatment of (pre-

viously) infected area with antibiotic ointment.

Myiasis (A & R)
Cause: Fly maggots, e.g., *Lucilla bufonivora* and others.
Symptoms: Tissue damage, especially in the region of the eyes, nose, and cloaca.
Treatment: Mechanical removal with subsequent treatment with cod liver oil ointment.

Worm Infestations (A & R)

Tapeworms (Cestoda): Proteocephala, Pseudophyllidae, Cyclophyllidae, etc.
Symptoms: Presence determined through fecal samples (segments, eggs), general weakening of host, mature worms in digestive tract. Second larval stage of Pseudophyllidae (plerocercoid) found below skin or in musculature and often externally visible by the presence of nodules or lesions (boils).
Treatment: (*) Adults in digestive tract: niclosamide (150-200 mg/kg BW) orally or Droncit® (5 mg/kg BW against Cyclophyllidae and Proteophyllidae; against Pseudophyllidae, 25 mg/kg BW). Plerocercoid: manual removal (surgically), followed by treatment of wound with antibiotic ointment.

Roundworms (Nematoda): ascarids, strongylids, spirurids, oxyurids of digestive tract
Symptoms: Can usually be determined from fecal samples (eggs, larvae, worms); weakening of host, tissue damage, tumor formation.
Treatment: (*) Panacur® (30-50 mg/kg BW), tetramisole (150-300 mg/kg BW), Fenbendozal® (30-50 mg/kg BW), or Rintal® (10-30 mg/kg BW) given orally. With small amphibians tetramisole is added to the bathing water.

Rhabdiasid roundworms of the lung
Symptoms: Pneumonia, tissue damage; free-moving first larval stage often found in mucus sample from oral cavity or in fresh droppings.
Treatment: (*) Citarin-L® (50 mg/kg BW) diluted and injected subcutaneously.

Dracunculid roundworms in tissue; filiarid roundworms in tissue and blood vessels

Symptoms: Extent of damage dependent upon numbers and sites of parasites; often loss of activity (reluctant to move about), weakness. Microfilarids from ovoviparous filarids found in bloodstream.
Treatment: Unknown.

Acanthocephalans
Symptoms: Infections of the digestive tract.
Treatment: (*) Niclosamide (150-200 mg/kg BW) 2 times orally.

Flukes (Trematodes)
Symptoms: Tissue damage of digestive tract and related organs.
Treatment: Difficult since liver damage can occur. Chloroquine (30-50 mg/kg BW) given twice orally or intramuscular injection.

III. VITAMIN AND MINERAL DEFICIENCY DISEASES

Rickets or rachitis (A & R)
Cause: Disturbance of calcium and phosphorus metabolism due to vitamin- and mineral-deficient nutrition.
Symptoms: Softening of bones and shell (turtles); vertebral and bone deformation; paralysis (especially in young animals).
Treatment: Vitamin D3 supplements together with calcium and mineral supplements or cuttlebone in diet. (*) Injection of multivitamin preparation with vitamin D3. Prophylactic: diversified vitamin-rich foods with mineral supplements. In addition (for reptiles), twice weekly UV radiation for 5 minutes or unfiltered sunlight.

Thyroid gland disturbance (R)
Cause: Iodine deficiency.
Symptoms: Swelling of crop (goiter) in lizards and turtles.
Treatment: Supplement drinking water with iodine.

Vitamin A Deficiency (R)
Cause: Lack of vitamin A.

Symptoms: Lidedema, infections of Harder's gland (of the eyelid) in turtles.

Treatment: (★) Three injections of aqueous multivitamin combination at intervals of 10 to 14 days; supplement food with cod liver oil.

IV. OTHER DISEASES

Intestinal Protrusion (A & R)

Cause: Presumably caused by incorrect maintenance, such as extremely dirty terrarium, overcrowding; sometimes due to intestinal nematode infestation (A).

Symptoms: Large intestine protrudes from cloaca, often followed by infections.

Treatment: Gently push back intestine with oily cotton wool; (★) surgical removal by veterinarian.

Fatty liver (A & R)

Cause: Presumably due to insecticide residues, viral infection, and other causes.

Symptoms: Detectable in tissue sections only.

Treatment: Unknown.

Difficulties in shedding skin (A & R)

Cause: Incorrect maintenance (usually kept too dry).

Symptoms: Incomplete shedding.

Treatment: Forced, prolonged baths (for several hours) in lukewarm fresh water, possibly with camomile additive; mechanical help to remove skin.

Fractures (A & R)

Cause: Accidents; incorrect handling.

Symptoms: Broken ribs and other injuries such as caused by incorrect catching of chameleons; shell fractures in turtles.

Treatment: Usually bone regeneration takes place. In order to prevent infections, antibiotic powder should be applied to a shell fracture in turtles.

Newt disease (A)

Cause: Presumably a nutritional deficiency.

Symptoms: Open, often "weeping" lesions and hemorrhaging sores.

Treatment: Bathing in potassium permanganate solution; in serious cases, amputation of the affected limb(s).

Viral scab disease (A)
Cause: Caused by virus (e.g., in *Lacerta viridis*).
Symptoms: Skin growth (papillomas and others); brownish black changes in appearance of the skin.
Treatment: (*) Surgical removal.

Poisoning (External) (A)
Cause: Overcrowding of terrarium.
Symptoms: Skin secretions from among the same or other species (e.g., among dendrobatids) can lead to cramps and paralysis symptoms.
Treatment: Sometimes immediate bathing in fresh water, possibly with some camomile, is effective. Transfer into a larger, clean terrarium.

Injuries (A & R)
Cause: Incorrect handling; fights.
Symptoms: Bites, skin damage, bruises, and other mechanical damage.
Treatment: (*) Amphibians (light cases): applications with or bathing in a diluted potassium permanganate solution. Amphibians and reptiles (serious cases): sulfonamid ointment or powder.

Constipation (A & R)
Cause: Lack of activity/movement; incorrect diet; many other possible causes.
Symptoms: Reluctance to move; no droppings.
Treatment: Bathing in lukewarm water for about 10 minutes; possibly enema application.

Visceral gout (R)
Cause: Presumably environmental temperature too high, with insufficient relative humidity; kidney malfunctions and other causes possible.
Symptoms: Uric acid deposits on serous skin, blood vessel walls, or organs; can only be determined from histological sections.
Treatment: Unknown.

Breeding and Rearing

INTRODUCTION

In this day and age of environmental awareness, any conscientious hobbyist will attempt not only to look after his animals correctly and in the best possible way, but he will also try to breed them. However, in order to accomplish that he has to understand the specific requirements for each species as well as the fundamentals of their reproductive biology. In addition, he must have patience and—if need be—also possess well-established, productive food animal breeding colonies.

The environmental climate with its abiotic factors of temperature, humidity, and light acts in nature as a time-related stimulant to initiate the reproductive cycle in poikilothermic animals. Amphibians and reptiles of temperate zones with their considerable temperature extremes between winter and summer will produce germ cells only during a short interval in spring, following the long winter rest period.

Consequently, they should also be given an identical opportunity in captivity. That is, they should be kept—at least for a few weeks, according to their natural habitat—in a cool, frost-free room to stimulate a winter rest period (please refer to winter rest in the chapter on housing).

Animals from a subtropical climate have a more extended reproductive period but usually also require a winter rest period to facilitate maturation of reproductive cells. For these animals it may be sufficient to turn off the lights and heating for a few weeks.

Animals from the tropics, especially those from areas with well-defined rainy and dry seasons, display a seasonal reproductive rhythm that is dependent upon the degree of humidity. Reproduction nearly always begins after the onset of the rainy season. In a terrarium this condition can be simulated by increasing the humidity (following a period of very low humidity) by setting up a continuous water misting or spray (sprinkler) system. For aquatic animals it is usually sufficient to raise the water level. Breeding of animals from equatorial and tropical zones, with their unchanging seasonal environmental conditions, can take place throughout the year.

Apart from climatic factors, the following points must also be taken into consideration:

Food: Should be varied and vitamin- and mineral-enriched. Excessive and insufficient feeding must be avoided.

Size of cage and decoration: Amphibians and reptiles display their natural reproductive behavior only in cages that are sufficiently large. Suitable spawning or egg deposition sites must be provided in the form of adequate water space, plants, caves, or substrate.

Stocking density: Overcrowding must be avoided. In those amphibian and reptile species that in nature are solitary except during the breeding season, males and females should be kept separately until breeding is to take place. This is the best method to assure breeding success. Territorial species in small terraria should only be kept in pairs.

Correct pairing off: If one discovers that all specimens from a group of young raised animals are of the same sex, or if—following correct husbandry procedures with mated pairs—breeding does not take place, the adult animals should be exchanged through contacts with other hobbyists, clubs, or breeding groups.

Proper record-keeping is an integral part of any animal breeding program. It should commence at the onset of breeding activities, and all significant events should be recorded. This facilitates reproducing particular procedures if necessary, or the records and information can be disseminated in the form of published articles. Important points to be recorded include:

   . . . age of adults, date, time, temperature, and humidity at the time of mating;

   . . . date when eggs were laid or date of birth in live-bearing species;

   . . . site of eggs/clutch deposition or number of juveniles per pregnancy;

   . . . egg size and—for reptiles—weight of eggs;

   . . . duration of embryonic development from egg to hatching of larva in amphibians or time of hatching of juve-

58

nile in reptiles and the larval development in amphibians from larva to metamorphosing juvenile with relevant temperature data;

   . . . dates of larval development (external gills, internal gills, front limbs, hind limbs);

   . . . first time food is taken;

   . . . appearance, weight, and size of juveniles (juvenile coloration);

   . . . time until sexual maturation.

Any disturbances—such as changes in the types of animals in the terrarium, changes of the terrarium setup, or possible transfer of pregnant or brood-caring animals—have to be absolutely avoided right from the onset of reproductive activities, since this can lead to total breeding failure. There should be no need here to elaborate on the fact that only healthy animals are used for breeding purposes.

AMPHIBIANS

EXTERNAL SEXUAL CHARACTERISTICS

The composition of breeding groups starts with the selection of suitable partners. Unfortunately, however, they can often be identified only on the basis of specific sex-related behavior during the mating season. Therefore, it is recommended that several specimens of the same species be kept together until the mating season. If sexual characteristics are well-defined in a species, they usually occur only at sexual maturity. It is therefore best to study and compare several animals during the breeding season.

The following secondary sexual characteristics often facilitate determination of males and females.

Newts and Salamanaders (Urodeles)

Body proportions: Females often stouter than males (e.g., *Cryptobranchus, Triturus*).

Cloacal region: During the breeding season this region is often swollen in males (e.g., in many ambystomatids and salamandrids).

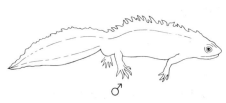

In newts and many salamanders, males ready to breed differ from females by having a greatly enlarged cloaca and often a distinct crest on the tail; a few European newts may even have a crest or comb along the back in the breeding male.

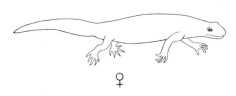

Body appendages: During the breeding season some male newts have serrated dorsal comb, dorsal fringes, or thread-like extended tail end (e.g., *Triturus, Notophthalmus, Cynops*).

Jaw (hedonic) gland: present in some male plethodontids.

Horny structures (pads) along the insides of thighs and toes as well as wider forelimbs or hind limbs in males of many salamandrids.

Frogs and Toads (Anurans)

Body proportions: Females larger than males (e.g., many bufonids, hylids, leptodactylids, rhacophorids, and ranids).

Cloacal region: Females with skin folds around cloacal region strongly swollen during breeding season (e.g., pipids).

Finger, lower arm, mouth region: During the breeding season males may have dark, horny courtship pads on fingers (e.g., many species *Bufo, Rana, Chiromantis, Litoria, Bombina*) and along the lower arm region (e.g., *Xenopus*) and/or along the oral margin (e.g., *Heleophryne*) or a dark thorn next to the first finger (e.g., *Leptodactylus*).

Upper arm broadened through presence of upper arm glands in males of *Pelobates*.

Throat region: When resting, throat skin in males colored darker (e.g., many hylids, hyperolids, ranids, microhylids). Slot-shaped gular fold present in males of *Phyllobates femoralis, Bufo maculatus, Ptychadena, Phrynobatrachus*, hylids, ranids, hyperolids, dendrobatids, bufonids, rhacophorids, microhylids, leptodactylids.

External sex characters in breeding *Xenopus laevis*. Males have patches of horny skin along the lower arms that are absent in females. Females have about three distinct lobes of skin above the cloaca.

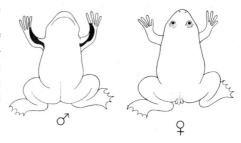

In *Pipa carvalhoi*, the three-clawed star-fingered toad, the female has a swollen ring of tissue around the cloaca when ready to breed.

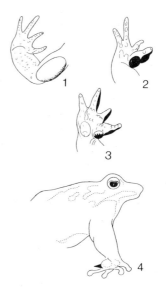

Male sexual characters in some frogs and toads. **1)** *Pelobates fuscus:* during the breeding season males develop an enlarged gland on the upper arm. **2)** During the breeding season *Rana temporaria* develops a dark swelling on the upper side of the thumb. **3)** *Bufo bufo:* during the breeding season males develop dark swellings on the undersides of the fingers and thumb. **4)** *Leptodactylus fallax:* during the breeding season males develop a horny black spur on a swelling above the first digit. (1-3 after Mertens, 1968.)

REPRODUCTIVE STRATEGIES

Amphibians have, in the course of their evolutionary development, developed a variety of reproductive strategies in order to adapt to their respective ecological niche. In this manner they have survived to now. Two main forms of reproductive strategies are clearly identifiable, and between the extremes on either side there are—as in all biological systems—smooth transitional forms. When applied to the various types of animal groups, they have to be viewed in a relative manner (as according to the socio-biologists Pianka and Wilson).

The *r-strategy* (reproductive strategy) takes advantage of favorable environmental conditions within an otherwise unstable and fluctuating environment by producing a large number of progeny in a short period of time. However, in a more stable environment the *k-strategy* (capacity strategy) is favored, whereby a smaller number of progeny is protected—usually by means of some sort of brood care or viviparity—against harmful environmental factors. So that this does not remain pure theory, here are a few practical examples. A typical r-strategist among the amphibians, such as the grass frog, is dependent upon the seasonal cycle and produces a maximum number of progeny during a short period of time in spring. However, the survival rate is severely affected since this phenomenon in turn fosters a massive occurrence of predators. Sudden climatic changes and deteriorating food availability place further restrictions on the survival rate.

On the other hand, in the habitat of poison arrow frogs in the tropical rain forests of Central and South America there is a relatively constant climate; moreover, the food supply remains largely constant and the mortality is low, especially since their skin poison renders these frogs less vulnerable to predation. In contrast to the grass frog, these anurans are therefore primarily k-selected; that is, their population size correlates with the capacity of a particular habitat. Survival of the species is assured despite a low reproductive rate because the frogs protect their young by means of special brood care mechanisms.

If one compares the forms of reproduction among the various species of Dendrobatidae, then one can recognize even here r-strategies and k-strategies. In the course of their evolution these frogs have developed increasingly more complex courtship and brood care behaviors, together with a relatively large to very small number of progeny, which indicates three distinct stages:

Stage 1: Includes *Colostethus* and *Phyllobates* species such as *C. trinitatis, C. inguinalis, C. palmatus, Ph. pulchripectus, Ph. pictus. Ph. anthonyi, Ph. tricolor, Ph. boulengeri, Ph. parvulus, Ph. trivittatus,* and *Ph. bassleri,* and *Dendrobates silverstoni.* It is characterized by a short courtship phase lasting a few hours, a simple courtship ceremony with head amplexus, as well as by brood care with defending and watering the clutch by the male (exception: *C. inguinalis,* brood care by female). The number of eggs per clutch is large. The male transports most of the larvae from one clutch to water during one trip. The tadpoles are herbivorous (plant-eating) and carnivorous (flesh-eating).

Stage 2: Contains *Phyllobates* species such as *Ph. femoralis, Ph. lugubris, Ph. vittatus, Ph. terribilis,* and *Ph. bicolor* and *Dendrobates* species such as *D. auratus, D. azureus, D. leucomelas, D. tinctorius,* and *D. truncatus.* They are distinguished from Stage 1 by having courtships that last up to several days, during which both partners are actively—primarily tactile—courting around each other. The clutch is usually not guarded, but the male keeps it wet. Upon hatching, the larvae are taken in small groups or individually to water holes. The clutch is of medium size, and the larvae are both herbivorous and carnivorous.

Stage 3: Includes three dendrobatid species complexes. They are referred to as "complexes" because there is still more or less uncertainty about the taxonomic position of these species. Their long courtship phase is characterized by additional visual signals being given, such "waving forelimb" or "stalking." The clutch contains only a few eggs. The brood care of these animals is virtually unique among vertebrate animals. The female has such good memory capacity that she can find the individually deposited larvae at the base of bromeliad leaves in order to feed them—throughout their entire developmental period—with food eggs specially produced by the female for her young.

According to our studies, such highly differentiated, active brood care might have developed as an intermediate link via the *D. quinquevittatus* complex. Usually here too the female feeds the individually deposited larvae with eggs produced solely for that purpose, but the rather aggressive tadpoles can also feed themselves without these eggs by feeding herbivorously, carnivorously, or on eggs from other species, without any time lost in their development. They thus form the transition to the food specialists, those larvae that feed solely on their own specific eggs, representing the *D. histrionicus* and *D. pumilio* complexes.

It is not possible here, of course, to discuss the entire spectrum of reproductive forms in amphibians. The table will convey a summary of the more significant ones. It will also show the number of young that can be expected, what spawning substrate is needed, whether parental animals take care of their eggs and larvae or whether the clutches or larvae can be removed from the terrarium for artificial rearing. It should be noted here that clutch size is species-specific, but it is also dependent upon age, size, and condition of the female.

FORMS OF REPRODUCTION IN AMPHIBIANS

Newts and Salamanders

External Fertilization, Egg-laying:
Clutch size: Large, usually in excess of 100 up to 1000.
Spawning Site: Water; bottom substrate, objects, self-dug pits.
Type of clutch: Strings of eggs in gelatinous cases; egg packages in gelatinous envelopes.
Brood care: None (exception: male of *Cryptobranchus* guards eggs).
Examples: Hynobius, Ranodon, Andrias, Cryptobranchus.

Internal Fertilization, Egg–laying (spermatophore storage possible):
Clutch size: Large; usually in excess of 100 up to 1000.
Spawning site: Water; bottom substrate, objects.
Type of clutch: Egg packages, individual eggs.
Brood care: None.
Examples: Ambystoma, Siredon, Taricha, Notophthalmus, Triturus, Cynops, Euproctus, Pseudotriton, Pleurodeles, Siren, Tylototriton.

Nesters, Egg-laying:
Clutch size: Medium; less than 100 eggs.
Spawning site: Water: bottom substrate, objects. Land: damp, dark sites at ground level, caves.
Type of clutch: Individual eggs (water & land).

Brood care: Female guards eggs (water); female guards eggs and keeps them moist (land).
Examples: Water: *Amphiuma, Necturus.*
Land: *Ensatina, Desmognathus, Aneides, Plethodon.*

Internal Fertilization, Live-bearing:
Litter size: Medium, less than 100.
Birth place: Water.
Type of litter: Larvae.
Brood care: Embryos protected in uterus of female against hazardous environmental influences; possibly also uterine feeding.
Example: Salamandra salamandra.

Nesters, Live-bearing:
Litter size: Small; usually less than 5.
Birth place: Land; same habitat as that of female.
Type of litter: Metamorphosed young.
Brood care: Embryos fed in uterus with atrophied egg cells.
Examples: Salamandra s. bernadozi, Salamandra atra, Mertensiella luschani.

Frogs and Toads

External Fertilization, Egg-laying:
Clutch size: Large, more than 200 up to several 1000.
Spawning site: Water; water surface, objects, bottom substrate.
Type of clutch: Egg "string," film, or lumps.
Brood care: None.
Examples: Rana, Bufo, Hyla, Hymenochirus, Discoglossus, Megophrys, Pelobates, Xenopus, Phrynomerus.

Bubble Nesters, Egg-laying:
Clutch size: Large, more than 200 up to several 1000.
Spawning site: Water surface, shoreline, caves, or vegetation close to water.
Type of clutch: Foam nest (bubble nest).
Brood care: None.
Examples: Rhacophorus, Chiromantis, Limnodynastes, Engystomopus, Physalaemus, Leptodactylus.

Leaf Nesters, Egg-laying:
Clutch size: Medium, usually less than 200.
Spawning site: Plants above surface of water (leaves sometimes folded over clutch).
Type of clutch: Lumps of eggs.
Brood care: None.
Examples: Hyla, Hyperolius, Agalychnis, Phyllomedusa, Dendrophryniscus, Afrixalus.

Pit Nesters, Egg-layers:
Clutch size: Medium, usually less than 200.
Spawning site: Pit in substrate or burrow (often self-dug) on land.
Type of clutch: Lumps of eggs.
Brood care: Guard and (often) water the eggs, in part by digging channels to the nearest water source.
Examples: Hyla faber, H. boans, H. rosenbergi, H. pardalis, Leptopelis, Hemisus, Breviceps, Eleutherodactylus, Arthrolepis, Anhydrophryne.

Egg String Male Care, Egg-layers:
Clutch size: Small, usually less than 60.
Spawning site: Male takes over egg string after spawning and fertilization.
Type of clutch: Egg strings.
Brood care: Male carries egg string wrapped around hind legs until larvae hatch.
Example: Alytes obstetricans.

Dorsal Pouch Care, Egg-layers:
Clutch size: Small, usually less than 60 (more in *Pipa*, but not all develop).
Spawning site: After fertilization, eggs taken by male onto back or placed into dorsal pouch of female.
Type of clutch: Individual eggs.
Brood care: Protected against detrimental external influences. Embryonal and in part larval development up to larvae ready to hatch or juvenile frogs in brood pouch on back of female.
Examples: Pipa, Fritziana goeldi, Hemiphractus, Gastrotheca.

Stomach Brooders:
Clutch size: Small.

Spawning site: Female swallows eggs after fertilization.
Type of clutch: Individual eggs.
Brood care: Protected against detrimental external influences through embryonic and larval development in stomach of female. Juvenile frogs are spat out.
Example: Rheobatrachus silus.

Throat Brooders:
Clutch size: Small.
Spawning site: Damp soil.
Type of clutch: Individual eggs.
Brood care: Clutch guarded by male until hatching, then takes larvae into well-vascularized throat sac. Spits out juvenile frogs after metamorphosis.
Example: Rhinoderma darwini.

Arrow Frogs, Type I:
Clutch size: Small.
Spawning site: Bromeliad leaf, dark ground area.
Type of clutch: Individual eggs.
Brood care: Either male or female (mostly) guards clutch; upon hatching, tadpoles are transported on back of one parent to water.
Examples: Colosthethus, Phyllobates, Dendrobates.

Arrow Frogs, Type II:
Other items as above.
Brood care: Female feeds individually deposited larvae until end of metamorphosis with eggs produced especially for feeding purposes.
Examples: Dendrobates pumilio-complex, *D. histrionicus*-complex, *D. quinquevittatus*-complex.

Internal Fertilization, Live-bearing:
Litter size: Small, usually less than 16.
Spawning site: Habitat of female.
Type of clutch: Fully developed toads.
Brood care: Protected against damaging external influences by undergoing embryonic and larval development up to juvenile frog in uterus of female. Developing stages feed on secretions from uterus wall.
Example: Nectophrynoides.

EGG DEVELOPMENT, HANDLING THE SPAWN, AND RAISING YOUNG

The clutches of those species that do not perform brood care should be transferred, including the surrounding substrate, into separate rearing containers once spawning has been completed. This is important since in many cases either the parents and/or any of the other terrarium occupants will consider the eggs and developing larvae as welcome prey and feed on them. Eating of eggs and larvae among brood-caring species can often be attributed to disturbing the parents or overcrowding. However, in some dendrobatids this can also represent a form of sexual interference.

If the eggs are deposited outside the water, they invariably require an oxygen-rich "steamy" atmosphere for proper development. They will usually become covered with fungus if totally submerged in water. For instance, according to our observations the spawn of *Hyla ebraccata*, *Agalychnis callidryas*, and *Phyllomedusa tomopterna* (unlike that of most dendrobatid species) must not be immersed in water when reared artificially. Therefore, the spawn including the surrounding substrate is transferred into a plastic container that is placed in a large bowl with water and the whole thing covered with a lid. The hatching larvae have to be able to slide from the spawning substrate into the water. Eggs covered with fungus are usually not fertilized. In order to prevent spreading of fungus to healthy eggs, the infected ones are carefully removed with forceps or a small spoon. It is also possible that healthy eggs and embryos may develop fungus; unfortunately, treatment with fungicides effective for fish diseases is rarely effective on amphibians.

Amphibians have always been popular scientific research animals, particularly in experimental embryology, since the embryonic development is easily observed in the relatively large yolk-rich eggs. Anyone with good vision or a mere hand-held magnifying glass can observe the developmental stages of many newts and frogs.

Depending upon the temperature, a furrow develops within a few hours after spawning. Over many subsequent cleavages the morula, a multicelled hollow sphere, develops. This is followed by the gastrula stage, which can first be recognized by the sickle-shaped blastopore and then by the clearly visible remnant yolk plug. The initially roundish egg is changed in shape by the following neurula stage. The medullary plate appears as the neural folds develop along the margin and close over into the neural tube. The small, elongated embryonic structure now becomes visible on the semispherical yolk sac. Within a few day this develops into a small larva that still has a relatively large yolk sac. Enzymes from the hatching gland dissolve the egg shell and the surrounding gelatinous mass, and then the tadpole emerges.

Until their yolk sac reserves are used up, young tadpoles use their adhesive pits around the head region to attach to plants or other objects in the water or to the container walls. After this stage they swim around actively searching for food. While newt larvae retain their external gills until the end of metamorphosis, in frog tadpoles these gills are displaced— before and also after free-swimming begins—to the interior of the oral cavity. Usually this is accompanied by the closure of one gill cleft, while the other one is transformed into a tube-like structure, the spiracle, that serves in inhalation and exhalation of water over the gills.

Once the larvae have left their gelatinous envelope they have to be split up into smaller groups and should be distributed over several containers with a variety of water plants and oak leaves as food and shelter. Ceramic tubes and similar objects can be added to provide other hiding places. Overcrowding can lead to excessive food competition, inhibited development, or increased intraspecific aggression.

Glass or plastic containers are suitable for groups of larvae that are non-aggressive or only mildly aggressive; this is recognizable by their predominantly herbivorous feeding habits. Strongly aggressive larvae (mostly omnivorous or carnivo-

A divided plastic container with individual compartments used for rearing specimens of some of the more aggressive amphibian larvae or for artificially rearing larvae of the *Dendrobates pumilio* and *D. histrionicus* complexes.

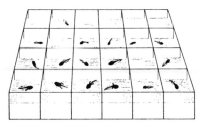

rous) or those that tend to inhibit each other's growth and development (*Dendrobates histrionicus, D. pumilio,* and similar species) have successfully been maintained in small refrigerator storage containers or subdivided containers made of inert (plastic) materials. The larvae of amphibians that spawn in flowing waters require clean, oxygen-rich water. In essence then, they should always be given an aquarium with a filter and aeration.

Despite even the best technical setup, water will have to be changed about once or twice a week, depending upon how dirty the tank is, the species involved, and the stocking density. Because of accumulating metabolic waste products and left-over food, such periodic cleaning is unavoidable. Feeding the larvae of some dendrobatids with egg yolk necessitates a complete water change about 5 hours after feeding. Water quality parameters have to be identical with those in the wild, and the water must not be chlorinated. Our Stuttgart tap water—with its water coming from Lake Constanz—has a pH of 7.9, a total hardness of 9.2 degrees, and a carbonate hardness of 7.0 degrees, and is suitable for many of the amphibians discussed later in the text.

In nature amphibian larvae have a diverse food supply available. Thus, when rearing them in captivity they must have a steady food supply; that is, they must be fed immediately once all previously-given food has been eaten. The larvae of newts and salamanders are exclusively carnivorous; that is, they require animal proteins. However, larvae of frogs are primarily omnivorous. They feed on a mixture of animal and plant proteins. Only a few tadpoles have become true food specialists, such as those of the *Dendrobates histrionicus* and *D. pumilio* complexes. These feed exclusively on species-specific food eggs, and only this type of food can be digested by the species concerned. Rearing the larvae artificially is only possible with a strict diet of egg yolk as a substitute food and by keeping them individually in small plastic containers.

Omnivorous larvae of other species should be given a diet that is as varied as possible. The summary provides details about the most important food items used to raise amphibian larvae in captivity. Specific requirements for particular species will be discussed in detail in the species descriptions later.

FOODS FOR AMPHIBIAN LARVAE

Egg-eaters (oophages):
Egg yolk as substitute food; eggs of some other amphibian species; species-specific food eggs.

Meat-eaters (carnivores):
Infusoria; various commercial preparations for feeding small fish fry; very small crustaceans (cyclops, daphnia, and others); mosquito larvae; newly born guppies; beef heart; newly born ("pinkies") dead mice; other amphibian larvae; dry (aquarium) fish foods with large meat content.

Plant-feeders (herbivores):
Algae; water plants; untreated lettuce and spinach; nettle powder; dry (aquarium) fish foods with large vegetable content.

Examining the digestive tract or the mouth region of a tadpole can show us the type of feeding used by the animal. For instance, carnivorous larvae and egg-eaters have a relatively short digestive tract (few spiral folds in the digestive tract) and a reduced number of intestinal villi in comparison to primarily herbivorous tadpoles of an identical stage of development.

With a variable diet, frequent feeding, regular water changes, and avoiding overcrowding, newly metamorphosed frogs usually do not develop stiff, weak, or degenerated front limbs, symptoms that are sometimes described as "match stick disease."

The end of metamorphosis is externally visible in newts and salamanders by the regression of the external gills into small vestiges and the functional development of front and hind limbs. In frog tadpoles the hind limbs develop first, even before the front limbs break through their skin pockets and before the tail is resorbed. Now the water level has to be lowered and one has to provide facilities for the metamorphosed young to climb onto land. This can easily be accomplished by placing the container at an angle or by providing cork or styrofoam floats. The young are then transferred to a rearing container. Terraria and commercially available plastic containers are suitable for that purpose when equipped with damp foam rubber, moss, or leaves as a bottom substrate. Additional hiding places should also be available, such as plants, cork pieces, or branches. These items have to conform to the normal requirements and satisfy the behavior of the animals concerned. Everything inside such a terrarium must be easy to maintain and be quickly cleanable.

Young frogs and salamanders require a large amount of small food such as springtails, ground mites, young leaf lice, or very small moth maggots. Some of the older and larger animals can already handle fruitflies, newly hatched crickets, whiteworms, tubifex, and other foods of suitable size. Some newly metamorphosed species can take food as large as houseflies. A broad vitamin- and mineral-rich diet, frequent feedings, and utmost cleanliness are mandatory requirements for successful amphibian breeding. The time required to reach sexual maturity is species-specific; but it is also dependent upon temperature, food supply, and stocking density.

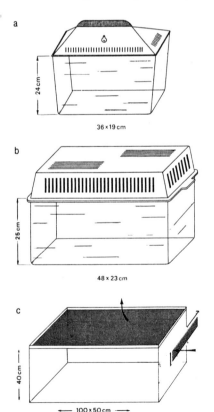

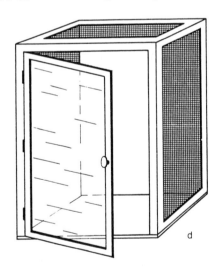

Rearing containers for young amphibians and reptiles. **a & b)** Clear plastic containers with lids. **c)** Frameless all-glass aquarium with wire mesh cover and sliding panel. **d)** Wire and glass insect box used for rearing young chameleons. After Zimmermann, 1982.

REPTILES

EXTERNAL SEXUAL CHARACTERISTICS

In reptiles, just as in amphibians, externally visible sexual characteristics can usually only be seen in sexually mature animals, primarily during the mating season. Even then they are often discernible only by comparisons among several specimens of the same species. The listing below gives a summary of the most important, although not always clearly distinctive, secondary sexual characteristics of reptiles.

Turtles and Tortoises

Body structure: Females of many turtles and tortoises are larger and heavier than males.

Plastron (abdominal shell): Often slightly concave near middle in males of many species.

Anterior limbs: With conspicuously lengthened claws in males of many turtle species, especially aquatic species.

Tail: Longer in males than in females; frequently thickened at the base and flattened and spindle-shaped.

Cloaca: Cloacal opening in male further from plastron than in females.

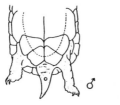 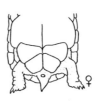

External sex characters in the tortoise *Testudo hermanni*. Males have the plastron distinctly concave, especially posteriorly, while it is almost flat in females. The tail of the male is also longer than that of the female, with the cloacal opening more posterior to the rear edge of the plastron. (After Adrian, 1980.)

Lizards and Skinks

Body structure: Males larger and more robust than females.

Base of tail: Broader in males and/or clearly set off from body (e.g., many chamaeleonids, lacertids, geckonids).

Coloration: Males when excited and especially during the mating season are more colorful than females (e.g., *Lac-*

erta agilis with green flanks, *Lacerta viridis* with blue throat).

Body appendages: Males with larger gular folds, throat sacs, nape and dorsal combs, helmets, horns, etc. (e.g., basiliscs, *Anolis,* some agamids, *Chamaeleo hoehneli, C. montium, C. jacksoni, C. johnstonei*).

Heel spur: Present in males of *Chamaeleo gracilis, C. dilepsis,* etc.

Anal (pre-anal) and thigh (femoral) pores, postanal scales, or inguinal scales more numerous or strongly developed in males (e.g., in many lacertids, *Anolis, Callisaurus, Holbrookia, Sceloporus,* agamids).

Snakes

Body structure: Males often more slender than females. Counts of ventral and caudal scales often sex-specific within a single species or subspecies.

Tail (cloaca to tip of tail): In males often longer than in females and often wider at the base because of the penes.

In boas and pythons male anal spurs comparatively elongated (spurs occur on each side of the cloaca).

REPRODUCTIVE STRATEGIES

In contrast to amphibians, reptiles have become totally independent of free water for their reproduction. They deposit their yolk-rich eggs, covered with a parchment-like or calcium shell, on dry land. Fertilization occurs in the oviduct of the female before the egg becomes encapsulated in its shell. It should be noted here that in some species spermatozoa

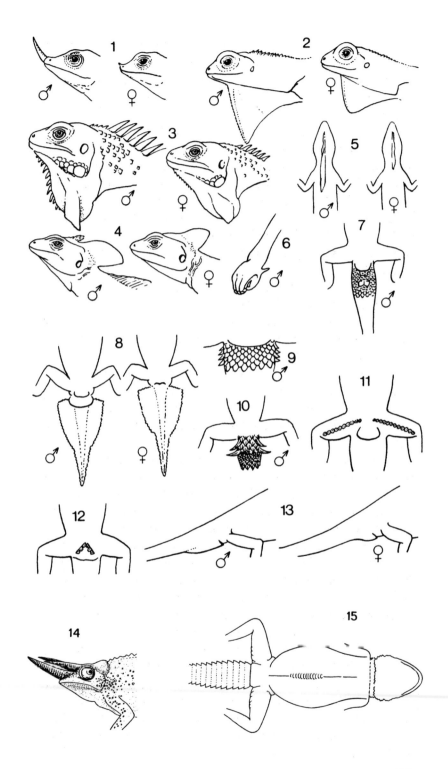

from one copulation can be stored for several years, allowing fertile clutches to be laid without new matings.

Through the development of different reproductive strategies, reptiles have been able to invade a variety of habitats, from the hottest deserts to cool, damp regions. Several families of lizards and snakes contain genera that are live-bearers. In these forms embryonic development up to fully developed juvenile takes place in the oviduct of the female, where the progeny is fully protected against detrimental habitat influences. While the young of most of these species feed on their large yolk sac, those of some skinks and the live-bearing geckos from New Zealand are fed by means of a placental connection with the maternal circulatory system. Those young surrounded by a thin shell without calcium tear it open during or shortly after birth and will then often feed on it as their first meal. This type of brood care is not only advantageous for the progeny, but the female also saves energy that she would have had to use in searching for a suitable place to deposit and bury the eggs.

Ovoviviparity or true viviparity occurs in *Lacerta vivipara, Chalcides, Tiliqua, Egernia, Anguis,* Cordylidae, some chamaeleonids, some non-venomous colubrid snakes, some venomous adders, most of the true vipers, and the boas, among others.

Asexual reproduction—parthenogenesis—provides another opportunity for a species to survive under unfavorable conditions. Only as recently as 1961 was this type of reproduction (the development of viable progeny from unfertilized eggs) first discovered by the Russian herpetologists Darewski and Kulikowa in a population of the *Lacerta saxicola*–group from the Caucasus. Since then at least another 27 lizard species have been discovered where this phenomenon occurs. These include: *Hemidactylus garnotii, Leiolepis triploida, Brookesia spectrum boulengeri, Cnemidophorus exsanguis,* and even a population of *Basiliscus basiliscus* from Colombia.

However, oviparity is the most common form of reproduction in reptiles, that is, the laying of fertilized undeveloped or only marginally developed eggs. This type of reproduction occurs in all tortoises and in most of the lizard and snake species. Only a few reptiles (e.g., some skinks and colubrid snakes, crocodiles, some adders, and pythons) also display some type of brood care.

DEPOSITION OF EGGS, INCUBATION, AND REARING THE YOUNG

Pregnant females have an increased demand for a vitamin- and calcium-rich diet as well as for an elevated liquid intake, demands that have to be catered to in captivity. Usually they stop feeding shortly before the eggs are laid and then begin to pace through the terrarium in search of a suitable place to deposit their eggs. We have to allow this to happen by providing a sufficiently large and deep substrate layer consisting of a loose, slightly damp mixture of peat moss and sand or green moss or by alternatively offering a suitable arrangement of plants, rocks, or branches and twigs.

For instance, the females of many geckos deposit their one or two eggs in cracks or crevices in rocks, branches, or cork pieces, on leaves, or even along the cage wall. These eggs, which are still soft for the first few hours, become solidly glued to the substrate they are deposited on. Soon their shells harden and the eggs then can not be detached again without damaging them. Most lizards, as well as turtles, bury their eggs in a moist, warm spot in loose soil. This is subsequently carefully smoothed over again so that there is barely a hint of the eggs being present. Snakes usually deposit their eggs in a casual manner in a protected, moist warm location, such as in or under rotting logs and debris. The individual eggs commonly stick to each other, often forming a clump of eggs.

In reptiles, size of the clutch is species-specific, but it also depends on the age, size, and condition of the female. If a pregnant female becomes egg-bound and subsequently dies, the reason for this could be improper care and maintenance, lack of a suitable place for depositing the eggs, or, among newly imported animals, poor condition and stress induced in transit. Should this occur there is a possibility of saving at least some of the progeny through Caesarean section. In a fresh-dead female, the abdominal cavity is carefully opened and the eggs are removed from the oviduct. However, these eggs must already be fertilized and ready to be laid. Using this method, we once managed to get 11 juveniles from eggs hatched in an incubator after they had been removed from a freshly imported *Amphibolurus barbatus* female that had died.

In most reptiles care for the progeny ceases after the eggs have been deposited in a suitable location. Thus, once a previously gravid female lizard or snake suddenly appears slim and emaciated, or when a female turtle returns to the water section after having spent a long time at one particular site on land, we have to start looking for the clutch. Although there is a possibility of raising the young in large, uncrowded terraria without human intervention, there is always a distinct possibility that the eggs or young will be eaten by other cage occupants, by food animals, or even by the parents. Therefore, it is often safer to look after the eggs. A fine-meshed wire "hood" can be placed over those eggs that cannot be removed from the substrate, but if possible, the eggs and the substrate they are attached to should be transferred to a special rearing cage. For that purpose buried eggs should be carefully unearthed and picked up. It is very important that the top of the egg is properly marked so that it is placed in the same position in the rearing cage; correct repositioning of eggs is important in order to prevent damage or death of the developing embryo.

We found the following rearing technique using a Type I brood container to be very effective for many lizards, snakes, and for some tortoise eggs:

A **Type I** brood container (incubator) useful for many lizard, snake, and turtle eggs. **1)** Bricks. **2)** Heater with thermostat. **3)** Thermometer. **4)** Wire insert to support egg container. **5)** Plastic container with substrate. **6)** Water. **7)** Glass panel placed at an angle to help retain humidity. **8)** Plastic container with lid.

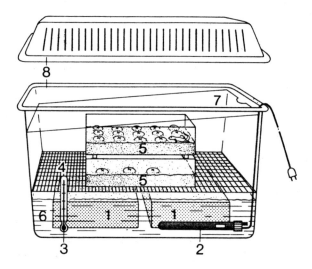

The marked eggs are each transferred into a slight depression in the slightly dampened substrate (which consists of a mixture of peat moss and sand largely sterilized through boiling) inside a 20 × 20 × 5 cm clear plastic container. Clutches with only a few eggs can be accommodated in smaller containers. These containers, which have a relative humidity of 90 to 100% (despite ventilating slots in the lid), are placed inside a plastic terrarium with the dimensions of 46 × 25 × 31 cm. The terrarium is filled with water to a level of 8 cm. There is a wire mesh insert about 1 cm above the water's surface. The water is then heated to the required temperature with a thermostatically controlled aquarium heater. The best temperature for most species is about 25 to 33° C. It must be remembered that soft-shelled eggs tend to increase in size and weight during their development through the intake of water from the environment.

Hard-shelled gecko eggs, which are less susceptible to drying out, are placed in a simpler rearing cage (Type II). The eggs of geckos from dry regions are placed in a depression in the sterilized substrate of dry sand. Eggs from geckos occurring in damp regions—mostly attached to some object—are transferred with their substrate to a layer of foam rubber that is kept constantly moist and placed inside a plastic container with a lid. This is then placed in a warm location such as on top of fluorescent lights. The temperature to be maintained should be 23 to 35° C.

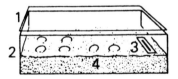 

Type II brood containers for hard-shelled gecko eggs from moist (left) and dry (right) regions. **1)** Lid with ventilation slits. **2)** Plastic container. **3)** Thermometer. **4)** Damp foam rubber (right) or suitable substrate (left). **5)** Eggs adhering to original substrate.

During the period of development, which is species-specific and temperature-dependent, the eggs have to be closely monitored. Unfertilized eggs and those with dead embryos become covered with fungus. They should be removed so as to not infect the remaining healthy eggs. There is another important point to be remembered: for some species the incubation temperature appears to be a sex-determining factor. For instance, in *Eublepharis macularius* it has been noted that with a constant temperature of 32 to 33.4° C mostly males will hatch from a particular clutch. On the other hand, a constant temperature of 26 to 29° C produces mainly females!

At the end of the incubation period the young will slice the egg shell open with an "egg tooth," a tiny horn-like elevation at the tip of the snout. In soft-shelled eggs imminent hatching is indicated by the appearance of small drops of liquid outside the shell together with a simultaneous shrinking of the shell. Soon thereafter the young pushes the tip of its snout out of the egg, but it still remains inside the egg for several hours. Finally, it will emerge through forceful rotational movements, thus breaking the shell open completely.

If a young reptile is too weak to release itself from the shell, this can be due to incorrect care, an insufficient vitamin and mineral supply in the diet of the parents, or incorrect (too high or too low) incubation temperatures. Should this occur, the young can be helped along if we cut cautiously into the egg shell or pull the shell carefully apart with two pairs of forceps. When doing this, attention has to be paid that the nostrils of the young are free immediately for normal respiration. Unfortunately, young saved this way almost invariably remain too weak to survive for long or they may have deformities.

Newly released young from egg-laying as well as live-bearing species are transferred in small groups into separate rearing tanks that have to conform to the species-specific environmental requirements. These terraria have to be set up in such a way that the animals are easily observed and the tank can be cleaned easily. There have to be adequate hiding and sunning places. Once the yolk sac remnants have been used up—usually a few days after hatching—the young have to find their first solid food. Turtles require various water plants, algae, or lettuce as rearing food, as well as daphnia, tiny shrimp, other small crustaceans, and mosquito larvae. This is then followed later on with earthworms, snails, small pieces of beef heart or raw fish, small whole fish (guppies!) or "pink" (newly born) mice. For aggressive species communal feeding may be difficult, leading to fights and injuries. In these cases, the animals should be kept separately or at least be fed individually.

Tortoises and herbivorous lizards should be offered a mixture of dandelion, clover, thistle, and other wildflowers free of insecticides and herbicides. These must be washed thoroughly and permitted to drip dry before feeding. Also often eagerly taken are small pieces of fruit and vegetables as available depending upon the growing season. These reptiles will also take small pieces of beef heart or newly born dead mice.

So that growing juveniles get sufficient vitamins and minerals, European tortoises and omnivorous lizard (including skinks) species should get a mixture of various baby foods supplemented with a few drops of multivitamin and mineral compounds, two to three times a week. The drinking water should also be fortified with calcium lactate.

Very small insect- and spider-feeding lizards should initially be fed on ample quantities of tiny arthropods such as springtails, leaf lice, and certain mites. Larger animals can get maggots of wax moths, mealworms, small crickets, beetle larvae, fruitflies, snails, or small pieces of beef heart. Young snakes usually will not take any food until after their first molt; then they will feed on small mice, strips of beef heart, chunks or strips of raw fish, and sometimes also insects. They have to be watched closely during the feeding, because it can happen that one juvenile snake will swallow not only its prey but also one of its siblings that has attacked the same prey. If a snake continues to refuse food for several weeks after molting for the first time, it must be force-fed.

In nature reptiles are always born at a time when there is usually an ample food supply. Therefore, we have to attempt to provide an equally diversified diet supplemented with vitamins and minerals. In addition, we must provide ultraviolet radiation to young diurnal reptiles by means of UV-A fluorescent tubes in continuous use or with 5-minute exposure to stronger UV sources at a distance of 1 m twice a week.

Groups of growing adolescent reptiles should not be kept too crowded; moreover, they must be constantly monitored. Only close supervision avoids stress mortalities. For instance, we have observed that the last-born *Amphibolurus barbatus* are often dominated by the first-born (6 to 10 animals on 2 square meters). Therefore, we have to hand-feed them outside the terrarium or place the food so that there is no visual contact with the more dominant specimens. Langewerf reports (personal communication) that with agamids he breeds, such as *A. stellio, A. caucasia, A. lehmanni,* and *A. sanguinolenta,* successful rearing occurs only when 20 animals are given an area of 3 square meters.

The time of sexual maturity in reptiles is species-specific, but it is also dependent upon food supply, temperature and stocking density.

The Terrarium Animals

PREAMBLE

The ancestors of modern amphibians developed from an off-shot of the bony fishes, the crossopterygians, during the Devonian, about 300 millions years ago. The development of walking appendages, stiffening of the vertebral column, and formation of lungs for atmospheric respiration enabled these animals to leave the water and explore new habitats on land. They reached their greatest diversification during the following Carboniferous and Permian Periods, but with the onset of the Mesozoic they were suppressed by reptilians, which were even better adapted to a terrestrial life.

To this day, most amphibians are still partially dependent upon water, at least for their reproductive cycle. Their moderately yolk-rich eggs are usually deposited in water, where the external layer swells up into a gelatinous mass. Embryonic development up to the metamorphosis stage—the transformation of aquatic, gill-breathing larvae into terrestrial animals—also takes place in water. Because of such aquatic development and their evolutionary relationships, amphibians occupy a somewhat intermediate position between fishes and truly terrestrial vertebrates such as reptiles, birds, and mammals.

As the first terrestrial vertebrates, reptiles were able to discover and utilize different terrestrial habitats with little competition. Formation of a horny body cover that can be differentiated into scales or plates provides effective protection against desiccation and external damage. Yolk-rich eggs and their parchment-like or calcareous shells, together with the formation of an embryonic sac (amnion) and a primary urinary sac (allantois), make the embryonic development of reptiles independent of an aquatic environment.

The first reptiles appeared about 260 million years ago, during the Carboniferous Period. The greatest species diversification—with the development of giant forms with body lengths of up to 26 m—was reached by the ancestors of today's reptiles during the Mesozoic Period. A large number of reptile groups became extinct toward the end of the Cretaceous Period. Because of their more efficient adaptive mechanisms, only the small and medium-size reptiles managed to survive and occupy ecological niches that could not be taken over by the ever stronger competition of warm-blooded birds and mammals.

The class Amphibia consists of three living orders: the legless caecilians (Apoda) with about 150 species; the newts and salamanders (Urodela) with over 300 species; and the frogs and toads (Anura) with about 2800 species.

The class Reptilia is divided into four recent orders: the turtles (Testudines) with about 230 species; the crocodiles (Crocodylia) with 21 species; the tuataras (Rhynchocephalia) with 1 species; and the scaly reptiles (Squamata) with suborders Sauria (lizards and skinks) with some 2900 species and the Serpentes (snakes) with about 2250 species. All together, reptiles and amphibians make up about 8500 species. To choose a representative selection from this species abundance is not only difficult but must always remain inherently incomplete. Therefore, the subsequent text will deal only with those species that are available either commercially or through exchange between hobbyists, AND where captive maintenance and breeding are possible. Typical representatives from each group, particularly in those cases where there is adequate specific information and detailed observation data available, are discussed in greater length. This is followed by specific details about species that can be kept and bred by the identical methods and procedures.

Crocodiles, venomous lizards and snakes, endangered species, and those species with highly specific food and maintenance requirements that are difficult to meet in captivity are not discussed. The listed terrarium animals also include those European amphibians and reptiles that have been bred in captivity.

AMPHIBIANS
Newts and Salamanders (Urodela)

Ambystoma mexicanum—Axolotl
(Family Ambystomatidae)
Description: Size up to 30 cm. Dorsum dark gray to black with lighter patches. Abdominal region light gray. Color mutants (piebalds, yellows, blacks, and true albinos, white with red eyes) raised in captivity. Three gill branches on each side of head. Males with enlarged lateral folds along sides of cloaca. Transformation of the normally neotenic (adults retaining larval characters) species into terrestrial animals is possible through administration of thyroid hormones. Frequently bred by hobbyists and as a laboratory animal. An endangered species in its natural habitat.
Distribution and habitat: Mexico, the canal system of former Lake Xochimilco.
Care: Feeds on worms, snails, mosquito larvae, small fishes such as guppies, and will also take small pieces of beef heart or other lean meat. Can be kept in an aquarium with fine gravel or sand as bottom substrate and water plants such as *Elodea, Myriophyllum,* or similar types. Provide weak to moderate aeration and filtration by means of an aquarium pump and filter. Water temperature 18-22° C. Can be kept as low as about 5° C during the winter months.
Behavior and breeding: In captivity the reproductive period of this neotenic salamander species extends over several months (November to June). At that time the male excretes an aromatic substance from its swollen cloacal region that attracts receptive females. The female approaches the male from behind and pushes her snout repeatedly against his cloacal region. This stimulates the male to release its gelatinous, pyramid-shaped spermatophore, which adheres to the rough bottom. The upper section of the pyramidal structure contains the whitish mass of sperm. The female maintains continuous snout contact with the cloacal region of the male, thus following him around in the tank. This forces the female to pass over the spermatophore, and she picks it up with her cloacal lips. This procedure is repeated several times; the male can deposit up to 25 spermatophores.

A few hours after the spermatophore transfer, which usually occurs during the hours of dusk or dawn, the female deposits 300 to 600 internally fertilized eggs individually or in clumps on water plants or simply on the bottom. Once this has been completed, the eggs should be removed from the aquarium by the keeper and then transferred into a container with a lower water level. The larvae will hatch in about 2 weeks at a temperature of approximately 20° C. They should be fed daily with brine shrimp, daphnia, and small tubifex or whiteworms. With a variable diet and clean water (filtered or, better yet, daily water changes) these animals can reach sexual maturity in about 1 year.

The following species can be kept and bred under virtually identical conditions:

Amphiuma means (Amphiumidae)—Two-toed Amphiuma

Maximum size: 116 cm.
Distribution: Southeastern USA.
Habitat: Acidic bodies of water, ditches, swamps.
Egg development and larval size: 20 weeks; 5.4 cm.
Remarks: Female guards egg string inside depression or pit; adults to be over-wintered at 15° C; should be kept in large aquarium.

Necturus maculosus (Proteidae)—Mudpuppy, Waterdog

Maximum size: 43 cm.
Distribution: Central and eastern USA.
Habitat: Bodies of water such as lakes, rivers, streams.
Egg development and larval size: 5 to 9 weeks; 2.2 cm.
Remarks: Females guard eggs laid individually under rock; sexual maturity in 4 to 6 years.

Siren intermedia (Sirenidae)—Lesser Siren
Maximum size: 69 cm.
Distribution: Southeastern and central USA.
Habitat: Calm, shallow bodies of water, such as lakes and swamps.
Egg development and larval size: 7 to 9 weeks; 1.1 cm.
Remarks: Sexual maturity in 2 years.

Ambystoma tigrinum—Tiger Salamander
(Family Ambystomatidae)

Description: Size up to 33 cm. Dorsum and abdomen olive with often irregular lighter yellowish patches, spots, or stripes; very variable, with many subspecies.

Distribution and habitat: Most of the USA south into Mexico. Found from lowlands into mountain ranges; underneath rocks and boulders, below leaves and fallen trees. Aquatic only during breeding season. Neoteny most common in some populations in ponds and lakes of western highlands.

Care: Takes worms, small snails, insects, and "moving" strips of raw meat. Should be kept in partially moist terrarium with half water and the other half a land section. The terrestrial end should be damp peat moss covered with moss or tree bark. The water section, with 10 to 15 cm water level, should have a sand substrate and a transition area to land with pebbles or large gravel. Water and air temperature about 20° C, and relative humidity 50% to 70%. During winter months as low as 5° C is possible.

Behavior and breeding: The courtship and reproductive behavior of this primarily terrestrial species take place at the bottom of lakes and ponds. The male applies a special strategy in order to remove a selected partner from among a group of other males and at the same time persuade her to pick up his spermatophores. When the male sees a potential partner (presumably the female is recognized by means of a species-specific odor) he swims toward her and then pushes her with head butting to the abdominal region and into her sides to an area in the water that is deemed safe and suitable. Occasionally the male's tail passes over the female's head. Once out of sight of the other conspecifics, the male stops this action and then begins to walk in front of the female, giving her gentle slaps with his tail against her head and back. The female follows the male, pushing with her snout continuously against the male's cloacal glands. Apparently this is the sign for the male to release his spermatophore, which is then ejected with a slightly raised and bent tail. The female then moves immediately over this sperm package and picks it up with her cloacal lips, moving it into her cloaca.

However, if the male has not been able to isolate his intended partner far enough away from the rest of the group of sexually mature males, it can happen that one of the other males moves between the partners just when the sperm package is transferred. This new male then covers the spermatophore from the male in front with his own body. Moreover, he also makes contact through his tail slaps with the female swimming behind. Now instead of the first male depositing his sperm package, the new male will release his spermatophore and so assure the perpetuation of his own genetic material. About 24 hours after spermatophore transfer the female will lay gradually up to 1400 eggs, either individually or in clumps, attached to water plants. The larvae develop in about 2 weeks, depending upon the water temperature. Rearing them is identical to rearing axolotls (separate tanks). Give daily feedings with brine shrimp, then mosquito larvae, daphnia, small tubifex and white worms; in addition, make daily water changes if possible. Once the external gills of the larvae begin to regress, the water level must be substantially lowered and a terrestrial area provided for the metamorphosed young salamanders. Rearing is just as in *Cynops pyrrhogaster*.

The following *Ambystoma* species exhibit similar courtship and breeding behavior and can be kept and bred under identical conditions:

Ambystoma cingulatum—**Flatwoods Salamander**
Maximum size: 13 cm.
Distribution: Southeastern USA.
Habitat: Pine forests.
Egg development and larval size: 3-5 weeks; 1.3 cm.
Larval life and juvenile size: 4 to 6 months; 7 cm.

Ambystoma gracile—**Northwestern Salamander**
Maximum size: 22 cm.
Distribution: Northwestern USA and British Columbia.
Habitat: Swampy and damp areas near water.
Egg development and larval size: 2 to 4 weeks; 1.6 cm.
Larval life and juvenile size: 12 to 24 months; 7.6 to 8.9 cm.

Ambystoma jeffersonianum—**Jefferson Salamander**
Maximum size: 21 cm.
Distribution: Northeastern USA and Canada.
Habitat: Forest in proximity of water.

Egg development and larval size: 4 to 6 weeks; 1.3 cm.
Larval life and juvenile size: 4 to 6 months; 5.1 cm to 7.6 cm.

Ambystoma macrodactylum—Long-toed Salamander
Maximum size: 17 cm.
Distribution: Northwestern USA and Canada.
Habitat: Dry to damp areas with ample vegetation.
Egg development and larval size: 2 to 6 weeks; 1.1 cm.
Larval life and juvenile size: 6 to 12 months; 4.8 to 9.8 cm.

Ambystoma maculatum—Spotted Salamander
Maximum size: 25 cm.
Distribution: Eastern USA.
Habitat: Open forests in proximity of water.
Egg development and larval size: 4 to 8 weeks; 1.3 cm.
Larval life and juvenile size: 2 to 4 months; 5 to 7 cm.

Ambystoma opacum—Marbled Salamander
Maximum size: 13 cm.
Distribution: Eastern USA.
Habitat: Forests.
Egg development and larval size: Courtship and egg deposition on land; female waters and guards clutch 2 to 8 weeks; larvae aquatic, 1.9 cm.
Larval life and juvenile size: 4 to 6 months; 7 cm. Sexual maturity at 15 months.

Cynops pyrrhogaster—Japanese Fire-bellied Newt
(Family Salamandridae)
Description: Size up to 14 cm. Dorsum black to brown, sometimes with red spots. Abdomen orange to red with black patches. Parotid glands along sides of head. During mating season has thread-like extended tail tip and enlarged cloaca; sides of body and tail bluish.
Distribution and habitat: Japan and eastern China. Standing bodies of water with heavy plant growth, maximum altitude 1500 m, mostly aquatic.
Care: Food: Daphnia, worms, snails, insects, small pieces of meat. Should be kept in aquarium with floating pieces of cork, tree bark, and/or rock structures with lush plant growth (*Myriophyllum, Elodea,* and others) or in an aquaterrarium with a large water section, about 20° C, with shaded portion.

Over-wintering in water or on moss at 5 to 10° C.

Behavior and breeding: During the courtship the male approaches the female and butts his head sideways against her body. If she continues to move forward the male swims around her in order to stop her. Positioning himself directly in front of the female, the male rubs his parotid glands along the female. At the same time the male's tail is used to "fan" chemical stimulants to the female to persuade her to pick up the spermatophores. The following transfer of the sperm package is similar to that described for *Triturus* later on.

The female deposits her eggs individually on water plants, and the edges of each leaf are folded over one egg. These leaves (with the eggs kept in place) can be cut off and transferred into a rearing container with clean (filtered) water so they cannot be eaten by other occupants in the same tank. Depending upon the temperature, the larvae hatch in about 10 to 12 days. They should be fed with cyclops, daphnia, and chopped worms. During metamorphosis the water level must be lowered and provisions made to enable them to climb onto land. Commercially available clear plastic containers padded with a layer of damp foam rubber are very suitable for raising juvenile newts. These containers can be easily monitored and quickly cleaned, and they provide the required high humidity. Some broken pottery or pieces of cork can serve as hiding places. The diet should consist of small worms, incapacitated fruitflies and their larvae, as well as other slow-moving insects. Four months after metamorphosis the young salamander can be returned to an aquarium. Sexual maturity is reached in 2 years.

The following primarily aquatic newt and salamander species can be bred under identical conditions:

Cynops ensicauda **(Salamandridae)**
Maximum size: 16 cm.
Distribution: Japan.
Habitat: Standing waters with substantial plant growth.
Egg development and larval size: 2-3 weeks; 1 cm.
Larval life and juvenile length: 3-5 months; 4.5 cm.

**Paramesotriton hongkongensis (Salamandridae)—
Hong Kong Newt**
Maximum size: 16 cm.
Distribution: Hong Kong.
Habitat: Flowing or standing waters with substantial plant growth.
Egg development and larval size: 4-6 weeks; 1 cm.
Larval life: 4 months.

Pseudotriton ruber (Plethodontidae)—Red Salamander
Maximum size: 18 cm.
Distribution: Eastern USA.
Habitat: Flowing waters. Adults in or near springs, etc., with clear, cool water.
Egg development and larval size: 6-10 weeks; 2.2 cm.
Larval life and juvenile size: 30 months; 11 cm.

**Pseudotriton montanus (Plethodontidae)—Mud
Salamander**
Maximum size: 20 cm.
Distribution: Eastern USA.
Habitat: Flowing waters. Adults in mud, crayfish burrows, etc., near or in cool water.
Egg development and larval size: 6-10 weeks; 1.9 cm.
Larval life and juvenile size: 18-30 months; 7.6 cm. Sexual maturity in 2½ years.

**Stereochilus marginatus (Plethodontidae)—Many-
lined Salamander**
Maximum size: 11 cm.
Distribution: Southeastern USA.
Habitat: Standing or slowly flowing waters.
Egg development and larval size: 8-14 weeks; 1.4 cm.
Larval life and juvenile size: 13-28 months; 7 cm. Sexual maturity in 3 years.

Desmognathus fuscus—Dusky Salamander
(Family Plethodontidae)
Description: Size to 14 cm. Dorsum with a broad reddish brown to gray zig-zag stripe with paired light spots alongside

it. Ventral region variable, often dark with darker marbled pattern. Male with submandibular gland; hind limbs distinctly longer than forelimbs.

Distribution and habitat: Northeastern USA south to Florida and west to Texas. A complex of similar species and subspecies that are hard to differentiate. Damp forest regions in and along shady creeks and streams, underneath rocks, leaves, or bark.

Care: Food: Worms, small slugs, insects, and small aquatic crustaceans. Should be kept in a moist terrarium at about 18° C with a shaded water section. If possible use flowing water (circulating pump) and sand as the bottom substrate of the water section. Land section with layers of peat moss, green moss, rocks, and tree bark. Over-wintering in water possible (2 to 6° C).

Behavior and breeding: The courtship behavior of this lungless species takes place on land. Initially the male touches the female with his snout and usually curls his body around that of the female. With his head downward, the male—bent around the female in a semicircle—rubs his submandibular (mental) gland back and forth over the head and back of the female. This action, together with submandibular secretions, appears to prepare the female for the spermatophore transfer. The male then walks in front of the female and she follows him, stroking his tail with her forelimbs. This appears to be the signal for him to release the spermatophores. The female picks them up into her cloaca while passing over them.

Six to eight weeks later the female moves to a damp, protected area underneath rocks or tree bark, and there she deposits 12 to 36 eggs that form a clump by adhering to each other with their gelatinous membranes. The female will care for the eggs. She wraps her body around this clump of eggs and sometimes pokes her head into the eggs. Although she leaves the eggs temporarily, she will always return to them. The 1.5-cm larvae hatch after 30 to 35 days. They will stay in the brood cavity for another 2 weeks, then they will actively move into the water. Only there can they develop their external gills. After 7 to 9 months, at a size of about 3.8 cm, transformation to the terrestrial form takes place. However, in some very similar species development is completed inside the brood cave. Rearing procedures as for *Cynops pyrrhogaster*. Sexual maturity is reached after about 3 years.

The following plethodontid salamanders are essentially terrestrial and can be kept and bred much like *Desmognathus fuscus*:

Desmognathus aeneus—Cherokee Dusky Salamander
Maximum size: 6 cm.
Distribution: Southeastern USA.
Habitat: Damp forests.
Egg development and larval size: 5-7 weeks; 1.1 cm.
Remarks: Aquatic larval stage lacking; sexual maturity in 2 years.

Desmognathus ochrophaeus—Mountain Dusky Salamander
Maximum size: 11 cm.
Distribution: Appalachian Mountains of eastern USA.
Habitat: Damp mountain forests, stream banks.
Egg development and larval size: 6-9 weeks; 1.4 cm.
Larval life and juvenile size: 7-8 months; 2.5 cm.
Remarks: Sexual maturity in 3-4 years.

Eurycea bislineata—Two-lined Salamander
Maximum size: 12 cm.
Distribution: Eastern USA.
Habitat: Stream banks, damp areas.
Egg development and larval size: 4-5 weeks; 1.3 cm.
Larval life and juvenile size: 12-36 months; 4.4 cm.
Remarks: Oviposition occurs in water; eggs laid on plants.

Eurycea longicauda—Long-tailed Salamander
Maximum size: 20 cm.
Distribution: Eastern USA.
Habitat: Stream banks, damp areas.
Egg development and larval size: 6-8 weeks; 1.9 cm.
Larval life and juvenile size: 3½-7 months; 4.1 cm.
Remarks: Sexual maturity in 1-2 years.

Hemidactylium scutatum—Four-toed Salamander
Maximum size: 10 cm.
Distribution: Eastern USA.
Habitat: Damp areas in forests, especially sphagnum bogs.
Egg development and larval size: 6-8 weeks; 1.3 cm.

Larval life and juvenile size: 1½ months; 2.2 cm.
Remarks: Sexual maturity in 2½ years.

Plethodon cinereus—**Red-backed Salamander**
(Family Plethodontidae)

Description: Size to 13 cm. Dorsum reddish brown to dark gray, with red to light gray dorsal band (can also be absent). Ventral side marbled, black and white or black and yellow. 18 to 20 costal grooves (lateral furrows). Males with submandibular gland.

Distribution and habitat: Eastern USA and Canada; sparse or absent in the southern USA. Forests and brushy areas, under rocks, tree roots, and bark. Purely terrestrial. Often abundant.

Care: Food: Pillbugs (isopods), insects, slugs, worms, spiders. Should be kept in a moist terrarium at 16 to 20° C with a small water section. Land section consisting of top soil/peat moss mixture with peat moss slabs, rocks, and bark. Spray regularly with water. Can be over-wintered at about 10° C.

Behavior and breeding: This lungless salamander also exhibits a complex courtship ceremony prior to spermatophore transfer. This in essence assures that sexual readiness in both partners is synchronized. If a male meets a female he pushes her with his snout, "sniffs" at her, and sometimes even bites her. This is then followed by the so-called "foot dance" of the male, where the legs are individually and sequentially raised and lowered. The male moves his snout along the female's body until it reaches her head. He then pushes his snout against the female's throat region, moves underneath it, and proceeds forward. The female then moves on top of the male, i.e., "tail straddling walk," first touching the tail of the male with her forelimbs, then her throat moves over the base of the tail. The two animals then continue walking while in body-tail contact. The male continuously turns back to his partner to rub his submandibular gland against her throat and mouth region. Then both animals proceed as before. After a while the male stops, his tail begins to tremble, and the base of the tail is raised slightly and one spermatophore is deposited. The male then moves his tail over to one side, moves forward a short distance, stops, and raises and lowers the hind limbs. The female then moves forward, pressed against the bottom, over the sperm package and pushes her cloacal region against it and picks it up with the cloacal lips. If another male attempts to interfere in these courtship ceremonies, the courting male bites the intruder and drives him off.

The eggs are internally fertilized, and the female attaches 3 to 11 eggs individually against the ceiling of a small depression under a rock, piece of bark, or decaying log. Then she guards the clutch. She curls up below the clutch and thus apparently provides an optimal environment for the development of the eggs. The entire larval development takes place inside the gelatinous membranes. The first juvenile salamanders hatch after about 2 to 3 months at a size of 2.2 cm. When given a diversified diet of small arthropods, the young will reach sexual maturity in about 1½ years.

The following plethodontids have similar reproductive and brood care behavior and can be kept and bred under identical conditions:

Plethodon glutinosus—Slimy Salamander
Maximum size: 21 cm.
Distribution: Eastern USA.
Habitat: Damp forest areas.
Incubation period and juvenile size: 8-12 weeks; 2.5 to 3 cm.
Remarks: Sexually mature in 3 years.

Plethodon welleri—Weller's Salamander
Maximum size: 9 cm.
Distribution: Eastern USA (small area in NE Tennessee, SW Virginia, NW North Carolina).
Habitat: Spruce forests on mountains.
Incubation period and juvenile size: 12 to 14 weeks; 1.9 to 2.2 cm.
Remarks: Eggs deposited among leaves.

Plethodon vehiculum—Western Red-backed Salamander
Maximum size: 12 cm.
Distribution: Western USA and Canada (SW British Columbia to NW Oregon).
Habitat: Damp forests.
Incubation period and juvenile size: 14-18 weeks; 2.2 cm.
Remarks: Sexually mature in 2½ years.

Aneides aeneus—Green Salamander
Maximum size: 14 cm.
Distribution: Northeastern USA (SW Pennsylvania to N Alabama).
Habitat: Rock crevices, mountains.
Incubation period and juvenile size: 12-13 weeks; 2.2 cm.
Remarks: Sexually mature in 3 years.

Aneides lugubris—Arboreal Salamander
Maximum size: 19 cm.
Distribution: Western USA (California).
Habitat: Coastal forests.
Incubation perod and juvenile size: 12 to 16 weeks; 2.5 to 3 cm.
Remarks: Bites and squeaks!! To be over-wintered at 16 degrees C.

Batrachoseps attenuatus—California Slender Salamander
Maximum size: 14 cm.
Distribution: Western USA (California).
Habitat: Forests; grassy plains of coastal region.
Incubation period and juvenile size: 12 to 14 weeks; 1.3 cm.

Ensatina eschscholtzi—Ensatina Salamander
Maximum size: 15 cm.
Distribution: Western USA (Washington to California).
Habitat: Deciduous and nondeciduous forests in coastal regions.
Incubation period and juvenile size: 12 to 14 weeks; 1.9 cm.
Remarks: Sexually mature in 2½ years. Tail breaks off easily!!

Pleurodeles waltl—Rough Spanish Newt
(Family Salamandridae)
Description: Size to 30 cm. Flattened, roundish head. Tail with skin fold. Skin with warts; tips of ribs sharp, projecting through skin. Dorsum ochreous yellow to olive green, ventrally lighter with dark patches. Laterally yellow to reddish along rib arches. Points of toes and lower tail margin yellow-orange. Males with enlarged upper limbs (courtship swellings along inside of limb) during breeding season.

Top: *Ambystoma maculatum*, the spotted salamander. Photo by B. Kahl. **Bottom:** *Ambystoma mexicanum*, the axolotl. Photo by K. Paysan.

Distribution and habitat: Southern and western areas of Iberian Peninsula and northwestern Africa. Standing water, reservoirs, cisterns, ponds, and lakes; underneath rocks and fallen tree trunks, branches, etc.

Care: Food: Worms, small aquatic crustaceans, insects, snails, small strips of raw meat. Should be kept in an aquarium with a water level of 25 to 30 cm, with thoroughly washed sand as the bottom substrate. Water plants such as *Myriophyllum* and *Elodea* can be added. Water temperature during the winter months 6 to 8 °C.

Behavior and breeding: This species breeds from spring through fall. The courtship takes place in water. The male approaches the female from behind and crawls under her until the upper side of his head presses against her throat region, making rubbing movements. Then the male reaches up with his stout swollen forelimbs to the forelimbs of the female. The two partners will move about in this "piggyback" position for hours. The female then pushes her snout against one side of the male, which then releases one of his forelimbs, bends his body, and deposits one spermatophore close to her snout. The male then leads the female around in a circle until she is positioned directly above the sperm package so that she can pick it up.

A month later the female attaches 200 to 1000 eggs surrounded with substantial gelatinous capsules in clumps or small clusters on rocks or water plants. These egg clusters can be removed and transferred to a separate container with fresh, clean water. Depending upon the temperature, the gilled larvae will hatch in 5 to 14 days. Initially they should be given very finely powdered dry food, followed by chopped tubifex, whiteworms, cyclops, brine shrimp nauplii, or daphnia. At 25°C in a large aquarium, metamorphosis occurs anywhere from the third to fifth month after hatching. From that point on the juveniles can be treated as described for *Ambystoma tigrinum*. This species becomes sexually mature about 6 months after metamorphosis.

Salamandra salamandra—Fire Salamander
(Family Salamandridae)

Description: Size to 29 cm. Dorsum marbled yellow-black to orange-black or with two yellow to orange longitudinal stripes against a black background. One large parotid gland on each side behind head and rows of glands along back and sides. Tail round in cross-section.

Distribution and habitat: Northwestern Africa and central and southern Europe to western Asia. Damp forests in hill country. Hides during the day under loose rocks and boulders, tree trunks, stumps, or in burrows.

Care: Food: Insects, spiders, worms, and slugs. Should be kept in a moist terrarium at 15 to 21°C or in an outdoor shaded enclosure. The bottom substrate should consist of a mixture of leafy soil/peat moss covered with green moss and pieces of bark. Provide a small water bowl. Temperature during winter months 5 to 8° C.

Behavior and breeding: This species is largely terrestrial as well as being territorial. The animals almost always return to their original burrow after their nocturnal hunt for food. Investigations have shown that this "homing instinct" is largely visually controlled. Fire salamanders are indeed able to distinguish spacial patterns fairly accurately. Moreover, this species reacts even to stationary prey. Recognition of stationary patterns is based upon continuous eye movements, so that even a stationary animal can—without turning its head— perceive far more of the world around it than had previously been assumed. This is also very important for close-up orientation in the natural environment. It has been observed that salamanders tend to orient themselves according to particular landmarks that they have memorized on their homeward-bound trips during their nightly excursions. Fire salamanders are also capable of long-distance orientation. For one thing, this enables females to always find the same spawning site for depositing the larvae. On the other hand, in this way both sexes can always find the same winter quarters again, year after year.

The fire salamander's bright body colors serve as a warning display. If attacked by a predator such as a snake, this salamander secretes a sticky, poisonous substance. The salamander is then immediately released and from then on the predator will avoid this salamander because of the negative experience in conjunction with recognizing the warning coloration.

Courtship behavior of the fire salamander resembles that of the Spanish newt; however, it occurs on land. The male attempts to slide underneath the female and wrap his fore-

Top: Young larvae of the tiger salamander, *Ambystoma tigrinum*. Photo by R. Zukal. **Bottom:** *Ambystoma tigrinum velasci*, a Mexican subspecies of the tiger salamander. Photo by Dr. S. A. Minton.

Top: *Aneides ferrus*, the clouded salamander. Photo by Dr. S. A. Minton. **Bottom:** *Cryptobranchus alleganiensis*, the hellbender. Photo by Dr. S. A. Minton.

limbs from underneath around those of the female. When he is successful he will carry his partner on his back for hours, continuously rubbing the base of his tail over the cloacal region of the female. Once the female also starts to move the posterior part of her body from one side to the other, the male will raise his head slightly and rub the throat region of the female. At the same time a spermatophore will be deposited by the male and pressed firmly against the bottom. He then raises his cloaca slightly and moves the posterior part of his body over to one side so that the cloacal region of the female is positioned over the sperm package, which is then picked up by the female's cloaca.

During the following spring or summer the female moves into a clear, oxygen-rich pond or lake. (Consequently, the terrarium has to contain a suitable water section.) The female positions herself at the edge of the water and then, while elevating her body and twisting her tail, up to 70 larvae with a length of 25 to 35 mm are discharged within a short interval. They already have four limbs and finely branched three-tiered external gills. These larvae can be easily distinguished from other European newts by the presence of light-colored patches at the bases of forelimbs and hind limbs.

The rearing aquarium does not need to be specially equipped or decorated to house the fire salamander. All that is required are some large rocks to provide hiding and escape places. Overcrowding with its accompanying problems of cannibalism, competition for food, and uneven growth of the larvae must be avoided. The larvae are fed with cyclops, daphnia, tubifex, chopped whiteworms, and similar types of food. Depending upon the water temperature, metamorphosis begins after 3 to 5 months. The external gills and the tail fin begin to regress, and the young salamanders begin to breathe atmospheric oxygen through their lungs. Now is the time for the water level to be lowered, and hauling-out areas for the 3- to 8-cm-long salamanders also have to be provided. The animals can then be transferred to a moist terrarium for further rearing. The food requirements are the same as for adults at that time. Sexual maturity is reached after 3 to 4 years.

It should be noted here that there are populations of this species in which females give birth to fully developed (1.8 to 3 cm) juvenile salamanders much as does the Alp salamander. In this case larval development is completed inside the female.

Taricha torosa—Yellow-bellied Newt, California Newt
(Family Salamandridae)

Description: Size to 18 cm. Dorsum black to reddish brown. Ventral area yellowish to orange. During the breeding season males have swollen cloacal regions and breeding pads on front and hind limbs.

Distribution and habitat: Coastal ranges of California, USA. Brush and forest regions, underneath bark, tree stumps, rocks. Moves into water for breeding only (December to May); some populations are permanently aquatic.

Care: Food: Worms, slugs, isopods, mosquito larvae, daphnia, small aquatic crustaceans, spiders, and amphibian larvae and eggs. Should be maintained in a moist terrarium at 18 to 22° C. The substrate should consist of a mixture of leafy soil and peat moss, layers of moss, or tree bark. A water section with sand as a substrate should be provided. Breeding occurs only in a large water section with a water level of from 25 to 30 cm. Water plants such as *Myriophyllum* or *Elodea* can be used in the water section, which during the winter months is kept at 8 to 12° C.

Behavior and breeding: During its terrestrial life this species displays a defense behavior that is similar to that of the masked salamander, *Salamandrina terdigitata* : the animal curls its tail into a semicircle above its back, raises its head, and exhibits to the aggressor the brightly colored orange throat and tail region.

During the breeding season *T. torosa* moves into the water, where courtship takes place. Research by American scientists has shown that the female at that time secretes substances from the skin that attract males. This mechanism also serves to distinguish males from females. Once a male has made visual contact with the female, he will swim above her and reaches around her shoulder girdle with his forelimbs and around her posterior region with his hind limbs. The two animals may maintain this position for hours, during which the male continuously rubs his cloacal region over the posterior dorsal area of the female and his submandibular gland moves over the nostrils of the female. If a rival male appears he will carry the female to a safe spot using a few strong rudder-like

Top: *Desmognathus fuscus*, the dusky salamander. Photo by J. Dodd. **Bottom:** *Eurycea longicauda*, long-tailed salamanders. Photo by B. Kahl.

Cynops pyrrhogaster, the Japanese fire-bellied newt. Photo by B. Kahl.

lashes of his tail.

This behavior appears to stimulate the female to pick up the spermatophore. After a while the male releases his hold and climbs over the female's head down to the ground. He stops not far from her snout and with a slightly raised tail deposits a sperm package. The female moves over the spermatophore and picks it up with her cloaca. Fertilization occurs internally.

The female deposits up to 30 eggs in clumps on water plants. The spawn should be transferred to a separate tank so that it will not be eaten by other newts in the same tank. The 1.1-cm-long larvae hatch after 50 to 60 days. They are fed on *Artemia* nauplii, cyclops, daphnia, and similar zooplankton. Metamorphosis is completed about 3 months after hatching. At that time the water level must be lowered and provisions made for the 5.1-cm juveniles to climb onto land. With a varied diet (for details see *Cynops pyrrhogaster*), this species reaches sexual maturity in 3 years.

The following species of *Taricha* have similar reproductive behavior and can be kept and bred under identical conditions:

Taricha granulosa—**Rough-skinned Newt**
Maximum size: 19 cm.
Distribution: Northwestern USA.
Habitat: Standing and flowing waters with lush plant growth, as well as adjacent land areas (forests, fields).
Egg development and larval size: 5 to 10 weeks; 1.2 cm.
Larval life and juvenile size: 4 to 12 months; 5-8 cm.

Taricha rivularis—**Red-bellied Newt**
Maximum size: 20 cm.
Distribution: Northern California.
Habitat: Flowing waters and adjacent coastal forests.
Egg development and larval size: 4 to 11 weeks; 1.1 cm.
Larval life and juvenile size: 4 to 6 months; 5.1 cm.

Triturus vulgaris—**European Common Newt**
(Family Salamandridae)
Description: Size to 11 cm. Dorsum yellowish brown to

brown, often with darker spots. Black lateral head bands present against a whitish to red background with dark spots. During the breeding season males have an undulating skin fold from the nape to the tip of tail and a lower tail fold that is orange with dark triangular patches and a band above with a bluish sheen. There are skin folds on the toes of the hind limbs.

Distribution and habitat: Europe (except southern France, southern Italy, Iberian Peninsula) to western Asia. Various damp environments in lowlands and highlands, including grazing areas. Prefers sunny, open waters for breeding.

Care: Food: Worms, daphnia, cyclops, insects. Can be kept in a moist terrarium at 12 to 22°C outside the breeding season. Bottom substrate should be a mixture of leafy soil and peat moss, tree bark, and green moss. A transition area to land with large pebbles or smaller stones should be provided. Breeds in aquaria with water plants such as *Elodea* and *Myriophyllum*. Acceptable water temperatures range from 12 to 18° C, during the winter months 2 to 8° C.

Behavior and breeding: Courtship takes place in water during the spring. Initially a male swims toward any other individual and checks its identity by means of smelling the cloacal region. When he has found a suitable female he moves toward her at an oblique angle and with a stalking gait. He then positions himself in front of her with an arched back and slaps the water with the most posterior part of his tail, so creating a water current toward the female with his vibrating tail tip. This carries to the female chemical stimulants given off by the male cloacal glands, chemicals that are apparently designed to stimulate the female to pick up the spermatophores. Should the female attempt to escape, the male continuously repeats these movements directly in front of her, sometimes even "hopping" back slightly.

When the female begins to follow the male, he then immediately turns around and starts to walk in front of her with his tail stretched out. He then moves his tail in an undulating fashion directly in front of the snout of the female. If she touches his tail tip, he bends his tail into an S-shape held as close to his body as possible, then raises the base of the tail and deposits a sperm package with his legs spread out laterally. Then he turns back to the female and sways from one side to the other. She follows him, touches his tail again, and

Top: *Notophthalmus viridescens dorsalis*, the broken-striped subspecies of the red-spotted newt. Photo by A. Norman. **Bottom:** *Notophthalmus viridescens viridescens*, the red-spotted newt, in its eft stage. Photo by Dr. S. A. Minton.

The red salamander, *Pseudotriton ruber*. Photo by B. Kahl.

is then pushed back so that her body touches the spermatophore. Once her cloaca makes contact with the sperm package it widens and the entire spermatophore is picked up.

During the following weeks the female deposits 100 to 150 eggs (about 5 mm in size) onto water plants. She folds the water plant leaves over each individual egg with her hind legs.

The eggs and the respective water plants are then transferred by the keeper into another aquarium. First cleavage occurs after about 5 to 8 hours at 20° C. Two weeks later the first larvae begin to hatch. They should be fed primarily with small daphnia, cyclops, and other zooplankton. Transformation of gill-breathing larvae to lung-breathing juvenile newts about 2 to 3 cm long occurs after about 2 to 3 months.

The young are then transferred to a rearing container where they are given springtails, leaf lice, other small arthropods, or small worms such as tubifex or whiteworms. Only after they have become sexually mature (in about 2 years) will this species return to water in order to breed.

The following aquatic salamandrids exhibit a similar breeding behavior and can be kept and bred under identical conditions:

Triturus alpestris—Alpine Newt
Maximum size: 11 cm.
Distribution: Central and eastern Europe.
Habitat: Standing and flowing waters and adjacent terrestrial areas (pastures, forests).
Egg development and larval size: 2 to 4 weeks; 0.8 cm.
Larval life and juvenile size: 3 months; 3 to 4 cm; sexual maturity in 2 years.

Triturus cristatus—Crested Newt
Maximum size: 18 cm.
Distribution: Central and eastern Europe and central Asia.
Habitat: Standing waters and adjacent areas (forests, pastures).
Egg development and larval size: 2 to 3 weeks; 1 cm.
Larval life and juvenile size: 3 months; 5 to 7 cm; sexual maturity in 2 years.

Triturus helveticus—Swiss Newt
Maximum size: 11 cm.
Distribution: Western Europe.
Habitat: Standing waters and adjacent areas (forests, pastures).
Egg development and larval size: 2 to 3 weeks; 0.8 cm.
Larval life and juvenile size: 4 months; 3 cm; sexual maturity in 2 years.

Triturus marmoratus—Marbled Newt
Maximum size: 16 cm.
Distribution: Iberian Peninsula to central France.
Habitat: Standing waters and adjacent areas (fields, forests).
Egg development and larval size: 2 to 3 weeks; 1 cm.
Larval life and juvenile size: 3 months; 4.3 to 7 cm.

Triturus vittatus—Striped Newt
Maximum size: 16 cm.
Distribution: Caucasus, Asia Minor, northern Israel.
Habitat: Waters with lush plant growth and surrounding areas.
Egg development and larval size: 2 to 4 weeks; 1 cm.
Larval life and juvenile size: 3 months; 4.5 cm.

Notophthalmus viridescens—Red-spotted Newt
Maximum size: 10 cm.
Distribution: Eastern and central USA.
Habitat: Waters with lush plant growth and adjacent areas.
Egg development and larval size: 3 to 8 weeks; 0.8 cm.
Larval life and juvenile size: 3 months; 3 to 4 cm; following metamorphosis juveniles are usually terrestrial (called efts) or rarely aquatic; sexual maturity in 2½ years; adults are primarily aquatic.

Tylototriton verrucosus—Rough-skinned Crocodile Newt
(Family Salamandridae)
Description: Size to 18 cm. Blackish brown body. Limb areas and underside of tail orange. Thirteen to 14 knob-like glands along ribs. Parotid glands present; bony edges along outer margin of skull.
Distribution and habitat: Northern India to northern Thailand

Top: *Plethodon cinereus*, the lead-backed phase of the red-backed salamander. Photo by J. Dommers. **Bottom:** *Plethodon glutinosus*, the slimy salamander. Photo by J. Dommers.

Salamandra salamandra, fire salamanders.

in damp, shaded areas underneath rocks or decaying leaves.

Care: Food: Worms, slugs, insects, small pieces of beef heart. Should be kept in a moist terrarium at 20 to 25° C and a relative humidity of about 80%. Use fine gravel as a bottom substrate, with rocks, moss-covered bark, *Scindapsus*, or similar items. A small water section with a 4-cm water level increased to 20 cm during the breeding season should be provided. An aquarium with floating cork, rocky islands, or *Elodea* and other water plants can also be used. Reduce the temperature to 15°C for two months during winter.

Behavior and breeding: Reproductive behavior similar to that of *Pleurodeles waltl.* One to two months after the winter period, a male attempts to follow a female and tries to bring his hook-like arms, which are bent backwards, over those of the female. However, the female continues to pull loose and crawls back on land. These courting maneuvers can last for several days. When the female is finally willing, the male can then grab the female in such a way that their cloacas are firmly pressed against each other. Deposition of spermatophores could not be observed.

About 12 to 13 days later about 50 eggs are attached with individual threads to the roots of *Scindapsus*, moss, or floating pieces of cork. The eggs are either transferred to a rearing tank or the parents are removed to avoid predation.

At an air temperature of 21 to 23° C and a relative humidity of 100%, the 12-mm larvae hatch after 15 to 30 days. They should be distributed over several rearing tanks. An aquarium air pump must be used to provide adequate aeration. The first type of food for the larvae should be sifted daphnia. This is eventually followed by increasingly larger food items. Metamorphosis sets in after about 4½ months. The external gills and the skin folds regress, the skin becomes more granular, and the bony ridges begin to appear around the head. By then the young crocodile newts are about 6.5 to 7.3 cm long and can be transferred to terraria. For rearing details see *Cynops pyrrhogaster.*

AMPHIBIANS

Frogs and Toads (Anura)

Agalychnis callidryas—**Red-eyed Treefrog**
(Family Hylidae)
Description: Females to 7.5 cm, males somewhat smaller.
Dorsum green, sometimes with white spots. Sides blue to
brown, with white to yellowish crossbands. Abdominal region orange to red.
Distribution and habitat: Central America. Lowland rain forests and adjacent hill country. Usually found on trees covered
with epiphytes or lianas in the proximity of water.
Care: Food: Insects. Must be kept in a moist terrarium at 22
to 26° C during the day, lowered to 18 to 22° C at night. Relative humidity 60 to 90%; humidity level should be further
increased through frequent mistings in order to stimulate
breeding. The land section should consist of peat moss covered with green moss. Plants that may be used include *Scindapsus, Dieffenbachia, Philodendron,* and *Aglaonema.* This
species must have climbing branches. The water section
should be without obstructions or decoration.
Behavior and breeding: During the breeding season the males
of this nocturnal hylid species emit courtship calls (0.8 to 2.4
sec, 1.4 to 2.7 kHz) from plants above or near water. It appears that the males are stimulating each other by their vocalization. Thus, at least two males and one female (preferably
more) should be kept together. A female ready to mate is attracted to the calls; it approaches the caller and stops a few
centimeters in front of his snout. The male stops his calls,
approaches the female, and jumps on her back, where he
clasps his forelimbs around her pelvic region. With the much
smaller male on her back, the female now climbs through the
terrarium for several nights in search of a suitable spawning
site. She attaches several clutches of 11 to 78 eggs to large
leaves hanging over the water. The eggs are fertilized by the
male while the spawning procedure is still in progress.

The eggs are about 3 mm in diameter and are encased in a
gelatinous mass. They have a light green upper side and are
attached to the underside of a leaf. After one week the gelatin
envelope liquifies and the cluster of eggs, including 10-mm
larvae, slides into the water below.

Triturus helveticus, the Swiss newt. Photo by Juster.

Taricha torosa, the yellow-bellied or California newt, breeding male at bottom. Photo at top by A. Norman; that at bottom by B. Kahl.

The larvae should then be separated into several tanks that have some water plants. Water quality must be maintained with an aquarium filter and aerator. Upon resorption of the yolk sac, the young larvae are fed with finely crushed aquarium fish fry food, algae, and chemically untreated lettuce leaves. In contrast to tadpoles of other frog species, these larvae do not swim about along the bottom of the aquarium, but instead hang obliquely at the surface with their heads pointed at an angle of 45° to the surface and the tips of their tails vibrating slightly. Metamorphosis is completed in about 7 to 10 days, at which time the young frogs move to land or on top of floating pieces of cork. Once the young are transferred into a regular moist terrarium, they will feed at first on fruitflies and then quickly thereafter on small houseflies.

Bombina orientalis—Chinese Fire-bellied Toad
(Family Bombinidae)

Description: Size to 6 cm. Dorsum brown to light green with black and green patches. Ventral region marbled with red and black. During mating season males have courtship pads along the insides of the forelimbs. Captive-bred specimens often have yellow and black ventral markings. Crosses (hybridization) with *Bombina variegata* possible.

Distribution and habitat: Eastern Siberia to northeastern China and Korea; mountain streams.

Care: Food: Worms, insects, slugs, strips of beef heart, or raw fish. Can be kept in an aquarium at 23 to 25° C. Provide places for hauling out by putting large rocks, roots, etc., as islands. Water plants such as *Cyperus, Elodea, Scindapsus,* and others can be used. Can be kept outdoors during summer months if not too hot and dry.

Behavior and breeding: When threatened, this species (just like its relatives, the European fire-bellied toads) assumes a distinctive defensive posture, the "boat position." The back is bent downward, forming a concave surface and the limbs are turned against the body so that the red and black limb markings are visible. Sometimes they will even turn over and so warn off any real or potential enemy with their brightly colored abdominal areas. If these toads are actually attacked, glands distributed all along their back give off a poisonous white secretion.

Chinese fire-bellied toads are primarily nocturnal. During

the breeding season (usually in spring when in captivity), the male sounds off with drawn-out "unk unk" calls that are occasionally interrupted by "roo roo" sounds. If the female moves toward the caller, the male will jump on her back, grasping her around the pelvic region with his forelimbs. If the female signals her unwillingness by stretching out her hind limbs and making slight body vibrations, the male will release his grip. However, if the female remains calm and swims through the tank with the male on her back, spawning usually will occur during the following night. There may be up to 200 eggs that are attached individually or in clumps to plant roots, submerged leaves, or rocks. Once spawning has been completed, the 2 to 2.5 mm in diameter eggs should be transferred to an aquarium planted with *Elodea, Myriophyllum, Cabomba,* or similar types of plants.

The first tadpoles, about 7 mm long and gray in color, will hatch in 3 to 4 days. They adhere to the aquarium glass or water plants in a vertical position. After the yolk sac has been used up during the following few days, the larvae begin to swim around the tank in search of food. They should be given small amounts of finely crushed aquarium fry food or chopped tubifex several times daily. About 10 days later the tadpoles have reached a size of 38 to 40 mm and the hind limbs have become visible. Three weeks after hatching—at a size of about 44 mm—the forelimbs break through. Soon thereafter the first 13- to 15-mm toads will climb out of the water and onto floating pieces of cork or styrofoam islands. They still have a grayish black back and a dirty white ventral region. About 4 days after metamorphosis they will feed on *Drosophila,* small wax moths, leaf lice, and other small insects. Coloration begins to change after 3 months. Sexual maturity in this species is reached at an age of 12 months.

The following *Bombina* species exhibit a similar breeding behavior and can be kept and bred under the same conditions:

***Bombina bombina*—European Fire-bellied Toad**
Maximum size: 5 cm.
Distribution: Eastern Europe.

Top: *Triturus cristatus*, the crested newt, breeding male. Photo by K. Knaack.
Bottom: *Triturus alpestris*, the alpine newt. Photo by S. Frank.

The European common newt, *Triturus vulgaris*. **Right:** A breeding pair, the male above. Photo by B. Kahl. **Bottom:** A group of metamorphosing larvae. Photo by S. Damian.

Habitat: Shore areas of shallow water bodies in mountain regions.
Egg development and larval life: 3 to 4 days; 1½ months.
Juvenile size; remarks: 1.2 cm; to be kept at 4°C during winter months in damp moss.

Bombina variegata—Yellow-bellied Toad
Maximum size: 5 cm.
Distribution: Central and southern Europe.
Habitat: Shore areas of shallow bodies of water in lowlands.
Egg development and larval life: 3 to 4 days; 1½ months.
Juvenile size; remarks: 1.2 to 1.5 cm; to be kept at 4° C during winter months in damp moss.

Bufo blombergi—Blomberg's Toad, Colombian Giant Toad
(Family Bufonidae)
Description: Size to 25 cm. Dorsum reddish brown; sides dark brown; abdominal region white, often with dark patches. Male smaller than female, with dark brown courtship pads on the first digit.
Distribution and habitat: Colombia; damp forest regions.
Care: Food: Mice and other small mammals, insects, worms. Should be kept in moist terrarium at about 25° C with a land section (1/4 - 3/8 of terrarium) made up of peat moss, green moss, and pieces of cork. A small water section (1/8 - 3/8 of tank) should be maintained. Daily spraying with water is desirable.
Behavior and breeding: When harassed or attacked, Blomberg's toad displays a characteristic defensive behavior. It extends its forelimbs, raises its body off the ground, blows itself up with air, and then forces this air out again to make threatening clacking sounds. When strongly aroused this species even thrusts its head in the direction of the opponent. If an actual attack takes places, the animal discharges a poisonous sticky whitish substance from its large parotid glands.

This species is primarily nocturnal. Our males became sexually mature after 4 years, moved from land to water, and began calling for females. If a female approaches, the male grasps her around the shoulder region. This is followed quickly by spawning. With her back pushed in and the hind limbs partially extended, the female expels strands of eggs. During this time the male sits on her back, massaging the cloacal region of the female with his hind limbs and fertilizing the eggs. The eggs are whitish, about 1 to 2 mm in diameter, and there may be as many as 15,000 in long strands. They should be transferred into previously prepared aquaria equipped with a filter, air pump, and air stones. The larvae will hatch in 5 to 6 days at a temperature of 24° C. They are about 12 mm long. Until the yolk sac has been resorbed, the young hang from the gelatinous mass or the aquarium sides. Unfertilized eggs and dead, fungused larvae must be removed promptly. The tadpoles will start swimming freely in another day or so. They should then be fed with finely crushed aquarium fry foods. On the sixteenth day the hind limbs are distinguishable, and the forelimbs appear on the 36th day. The grayish brown patchy markings of the young toads now become visible and the tail regresses. At that point the water level must be reduced substantially and additional flotation devices such as pieces of cork, styrofoam, or foam rubber should be offered to enable the young toads to get out of the water and onto land. They will actually leave the water sometime between the 43rd to the 47th day.

The young toads should then be transferred to rearing cages that have a moist bottom substrate in the form of damp foam rubber and some hiding places such as pieces of cork. Initially they should be given a diet of small crickets, *Drosophila,* and mosquito larvae.

Bufo marinus (to 25 cm; Central and South America; forests and land under cultivation) can be kept and bred under the same conditions as Bufo blombergi. The eggs are deposited in long strands, and the larvae hatch within 3 days. Metamorphosis is completed in 6 to 7 weeks.

Bufo bufo—European Common Toad
(Family Bufonidae)
Description: Size to 15 cm, but varies geographically. Dorsum sandy colored to dark brown, frequently with darker spots and patches. Abdominal region whitish to gray, often with a dark marbled pattern. Iris golden to copper-colored. No external vocal sac. During breeding season males have black horny pads along the insides of the first three digits.
Distribution and habitat: Europe and northwestern Africa to Palearctic Asia and Japan. Found from lowlands into moun-

The aquatic flat-tailed caecilian, *Typhlonectes compressicauda*, is the only member of its group currently available. It has bred in captivity, females giving live birth to large larvae that resemble the adults. The male has the terminal cloaca modified into a copulatory organ (bottom photo) for internal fertilization. Photos by Dr. W. E. Burgess.

Agalychnis callidryas, the red-eyed treefrog. Photo by B. Kahl.

tain regions, in gardens, cultivated lands, pastures, vineyards, abandoned rock quarries, parks, and along the damp margins of forests. Found in standing waters during the mating season.

Care: Food: Insects, spiders, small slugs, worms. Can be kept either indoors or outdoors in a semi-moist terrarium at 18 to 26° C. The bottom substrate should consist of a slightly damp mixture of peat moss, sand, leaves, and some peat moss slabs. Additional hiding places must be provided such as small tree trunks, tree roots, or similar objects. There must also be a small water section. Use mainly ferns as decorative plants.

Behavior and breeding: Outside the breeding season this species is nocturnal and lives exclusively on land. It has a well-developed sense of orientation and during the breeding season it can relocate the water body in which it was hatched. Although they have a strong homing instinct that allows them to always return to the same breeding site, they can also be bred in captivity.

In the wild, male common toads will gather during the mating season (spring) at the edge of standing waters, where they will call for a female with muffled mating calls. Because of a still undefined innate triggering mechanism, each object the size of a mature female toad is approached and grasped. If this happens to be a male, he will identify himself as such through defensive squeaking and twitching lateral movements; he is then released. On the other hand, a female willing to mate tends to hold still, and with the male on her back she swims to an area with thick vegetation. She grasps a plant stem with her hind limbs and then assumes a signaling position that informs the male that spawning has begun.

Following several abdominal contractions, the female curls her back inward, lifts her head, and simultaneously extends her hind limbs laterally and toward the back. With his hind limbs pulled in and the surface areas of his feet pointing down, the male slides backwards, so that his hind limbs rest above the cloacal opening of the female, thus forming the so-called basket position. The male catches the double-stranded spawn in this position and fertilizes the eggs. The egg strands are 3 to 5 m long and can contain up to 10,000 eggs that are 1 to 1.5 mm in diameter. The strands usually are wrapped around submerged or semi-submerged water plants.

126

The first tadpoles will hatch after 12 to 18 days. In contrast to those of other species, these actually school; that is, they may form schools as wide as 1 m and several meters long swimming in large bodies of standing water. They are also eating simultaneously, usually detritus, algae, or (in an aquarium situation) dried aquarium fish food. If one of the larvae becomes injured it gives off a warning substance (pheromone) that stimulates flight among the siblings.

After 2 to 3 months (shorter in an aquarium situation due to the usually higher temperature) the small (0.8 to 1 cm) toads will leave the water. They can then be transferred to a semi-moist terrarium and fed with very small insects and similar foods.

Common toads have been used for neurobiological research. It was noted that the key stimulus "prey" is composed essentially of directional movement and area configuration. For instance, if an object moves perpendicular to the movement of the toad it will be considered as "prey," and thus initiate a turning maneuver toward the prey, sneaking up to it, fixing it in sight, and finally snapping it up. However, if a large object approaches at eye level it initiates either flight or (when directly threatened) sets up the defensive posture (inflation with air, stalking gait, and lateral body movements).

The following toads can be kept and bred under the same conditions:

***Bufo calamita*—Natterjack Toad**
Maximum size: 10 cm.
Distribution: Western and central Europe.
Habitat: Sandy areas.
Egg development and larval size: 3 to 4 days; 0.5 cm.
Larval life and juvenile size: 2 to 3 months; 1 cm; tadpoles live solitary; bottom substrate of terrarium should be sand, sexually mature in 2 years.

***Bufo viridis*—Green Toad**
Maximum size: 10 cm.
Distribution: Central and eastern Europe to northern Africa and central Asia.

Bombina orientalis, the Chinese fire-bellied toad. Photo by B. Kahl.

Top: *Bufo calamita*, the natterjack toad. Photo by G. Dibley. **Bottom:** Blomberg's toad, *Bufo blombergi*. Photo by H. Zimmermann.

Habitat: Dry, sandy lowlands.

Egg development and larval size: 3 to 4 days; 0.7 cm.

Larval life and juvenile size: 2 to 3 months; 1 cm; tadpoles live solitary; bottom substrate of terrarium should be sand; sexually mature in 2 years.

Chiromantis rufescens—Rough-skinned Foam-nest Treefrog

(Family Ranidae)

Description: Female to 6 cm, males about 1 cm smaller. Dorsum brown to grayish green with green dots. Abdominal region light green to white. Undersurface of hind limbs, as well as hand and foot areas, dark blue; tongue and below mouth blue.

Distribution and habitat: Western, central and eastern Africa. Tropical rain forests, on leaves and branches of trees and bushes and low vegetation such as *Lycopodium, Selaginella, Musanga, Harungana,* and *Hibiscus.*

Care: Food: Insects. Must be kept in a moist terrarium at 22 to 28° C. The land section should consist of damp peat moss covered with living moss. Climbing branches and large-leaved plants such as hibiscus should be provided. A small water section should be added, and all should be sprayed with water daily.

Behavior and breeding: This species is primarily nocturnal. During the day it sits motionless on branches and leaves with the extremities pulled close against the body. Because of its camouflage coloration, this frog is virtually indistinguishable from its surroundings. The breeding season coincides with the onset of the rainy season. This can be simulated in captivity by increasing the humidity level.

The males congregate among branches above seasonal streams and from there announce their mating willingness through communal calling. While doing this they also secrete a musk-like odor. When a female approaches, she will be immediately grasped around the shoulder region by one of the males; it is not uncommon to see additional males also climbing on top of the female by hanging on to either side of the first male already in copulation. Then the female excretes from her cloaca a light-colored viscous liquid that is formed inside a special "foam gland." This liquid is immediately beaten into a foam by swimming motions of the hind limbs of both male and female. Then the female expels the first eggs, which are immediately fertilized and pushed into the foam mass. Once spawning has been completed, additional foam is created to cover the eggs. In essence, a large foam nest is thus constructed by the spawning pair, which may have taken 1½ to 2 hours to spawn. Once this task has been accomplished, both male and female leave the spawning site.

The foam nest is about 10 × 5 × 6 cm in size and is suspended above water. The outer layer hardens by the next morning, while the inner nucleus remains moist and rubber-like. This thus protects the 110 to 200 whitish eggs (2.5 to 3 mm diameter with a thin gelatinous layer) against desiccation.

In about 5 to 8 days the eggs will give rise to 10 to 11 mm tadpoles with external gills. During subsequent heavy rainfalls (in a terrarium, following heavy spraying with water) the outside layer of the foam nest dissolves and the larvae fall into the water below. Their subsequent development takes place in water.

With a diet of algae, wheat germ, and commercially available dry tropical fish food, metamorphosis will be completed in about 10 weeks. Further rearing then should be done in a moist terrarium, where crickets, flies, and similar foods form the main diet.

The following foam-nesting frogs exhibit similar reproductive strategies and can be kept and bred in a slightly drier environment (semi-moist terrarium):

Chiromantis xerampelina
Maximum size: 9 cm.
Distribution: Southern and eastern Africa.
Habitat: Forest and bush regions, savannahs to rain forests.
Egg development and larval size: 5 to 6 days; 0.9 cm.
Juvenile size: 1.6 cm.

Chiromantis petersii
Maximum size: 9 cm.
Distribution: Eastern Africa.
Habitat: Grasslands and bush, savannahs.

Top: *Bufo carens*, the South African red toad. Photo by G. Dibley. **Bottom:** Territorial female of *Colostethus trinitatis*, a small arrow frog. In this species the throat of males is black, of territorial females yellow. Photo by H. Zimmermann.

Top: A territorial fight between *Dendrobates histrionicus* (on top) and *D. lehmanni* (on bottom) males. The *D. lehmanni* is the intruder in the territory of the *D. histrionicus*. Photo by H. Zimmermann. **Bottom:** *Dendrobates histrionicus*, a poison arrow frog. Photo by A. Norman.

Egg development and larval size: 4 days; 0.9 cm.
Larval life and juvenile size: 10 to 23 weeks; 2 to 3 cm.

Rhacophorus leucomystax (Rhacophoridae)
Maximum size: 7 cm.
Distribution: Eastern and southeastern Africa.
Habitat: Grasslands and bush, semi-moist and moist regions.
Egg development and larval size: 8 to 14 days; foam nest on the ground or above water.

Engystomops pustulosus (Leptodactylidae)
Maximum size: 3 cm.
Distribution: Central America.
Habitat: Proximity of water.
Egg development and larval size: 3 days; 0.4 cm.
Larval life and juvenile size; remarks: 2 months; 0.9 cm. Moist terrarium with large water section; 20 to 28°C; 50 to 150 eggs in foam nest at the edge of water. For young frogs use springtails as rearing food.

Pleurodema cinera (Leptodactylidae)
Maximum size: 4.6 cm.
Distribution: Central America to eastern South America.
Habitat: Proximity of flowing water.
Egg development and larval size: 2 days; 0.6 cm.
Larval life and juvenile size; remarks: 2 months; 1.3 cm. Care and maintenance as above; foam nest in water.

Dendrobates histrionicus & D. lehmanni—
Poison Arrow Frogs
(Family Dendrobatidae)
Description: Size to 3.8 cm. Base color black to brown with one or more yellow, orange, red, or white spots or bands, sometimes with reddish orange and black reticulated pattern on head and yellowish and black pattern over body.
Distribution and habitat: Colombia, Ecuador. Tropical lowlands and mountainous rain forests (18 to 1070 m elevation). On the ground and on fallen tree trunks, small shrubs, and bushes.
Care: Food: Insects such as small crickets, fruitflies and their larvae, wax moth maggots, leaf lice, etc. Should be kept in a

Dendrobates histrionicus **(h)** occurs in a multitude of colors and markings. On the basis of skin toxins, one color variant was separated from the rest as a distinct species, *D. lehmanni,* **(l)** in 1976. However, on the basis of similarities in social and reproductive behavior and larval biology, this species is not distinguishable from *D. histrionicus.* The hybrid progeny of these two "species" are fertile.

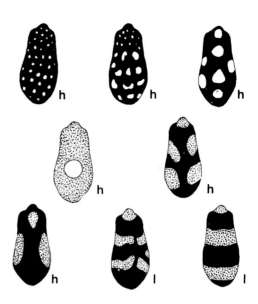

moist terrarium at 22 to 26° C; during the night keep at lower temperature of 18 to 22° C. Relative humidity 60 to 100%. Should be sprayed with water twice daily in morning and evening. For a substrate use clay soil covered with sheets of moss. Plants can include ferns, bromeliads, *Ficus, Scindapsus,* orchids, and similar plants. Spawning sites can be made from half coconut shells inverted on petri dishes or layers of leaves. Supply a water dish.

Behavior and breeding: These and all other dendrobatids listed in Stage 3 (see chapter on reproduction) occur in an enormous diversity of colors and markings, giving rise to many taxonomic inconsistencies and revisions. For instance, in 1976 *D. lehmanni,* the red-banded arrow frog, was split from *D. histrionicus* on the basis of difference in skin poisons. Despite a videoanalysis of our breeding groups of *D. histrionicus* and *D. lehmanni,* we could not detect any differences in the social, courtship, and brood care behavior or in the larval biology. Juvenile *D. lehmanni* occur in numerous marking variations. Hybridization between the two "species" is possible, and their progeny are fertile.

The males display the same kind of well-defined territorial behavior in a sufficiently large and suitably decorated terrarium as they do in the wild (this also applies to males of *D.*

Dendrobates lehmanni, the red-banded arrow frog, probably a variant of *D. histrionicus*. Photo by W. Mudrack.

Facing Page: Reproduction in the arrow frog *Dendrobates quinquevittatus*. **1)** Female. **2)** Male calling. **3)** Female joins male. **4)** Male and female together before laying. **5)** Eggs at three to four days of age. **6)** Newly hatched larvae with external gills. Photos by R. Bechter, courtesy Dr. D. Terver, Nancy Aquarium, France.

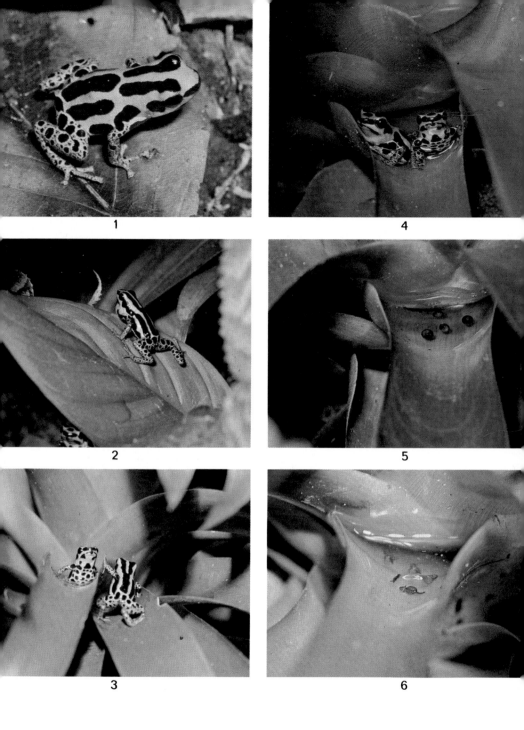

granuliferus, D. quinquevittatus, D. pumilio, and *D. speciosus).*
Transgressions of territorial borders while defending a territory as well as during the courtship of several males for the same female invariably lead to ritualized fights. A typical sequence between two males motivated to fight can be displayed as a sequential chain of reactions, as follows:

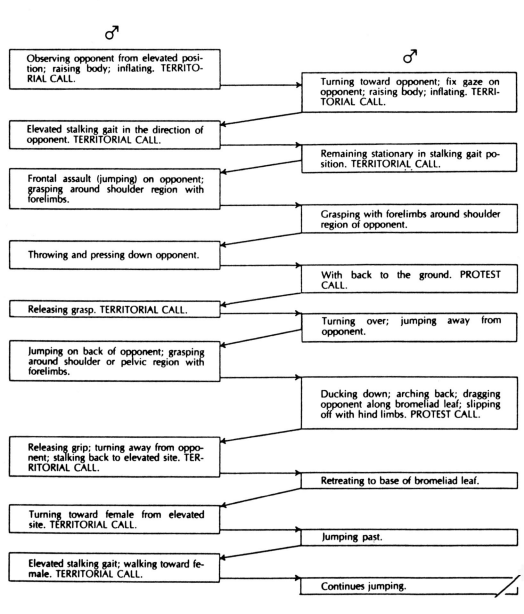

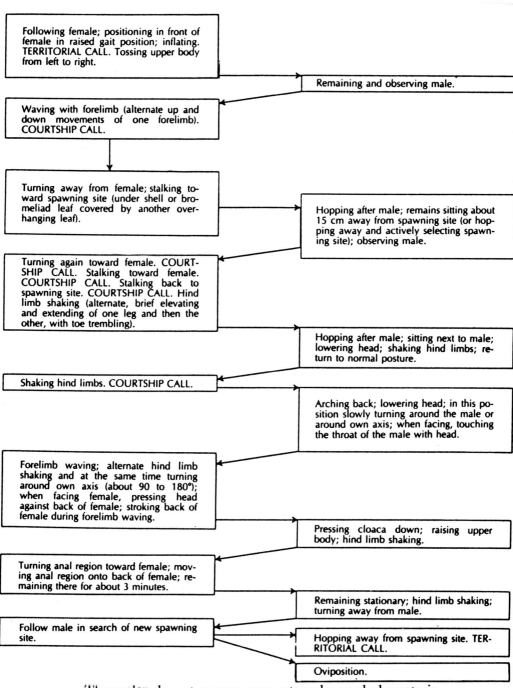

The males do not possess any external sexual characteristics. Therefore, it is important to be able to recognize them on the basis of their specific behavior so that proper pairs can be formed and losses due to stress avoided. Consequently, an

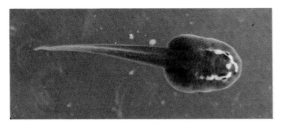

Dendrobates quinquevittatus. Photos by R. Bechter, courtesy Dr. D. Terver, Nancy Aquarium, France. **Top:** Tadpole showing beginning of yellow dorsal pattern. **Center:** Froglet before completing metamorphosis. **Bottom:** Male carrying a tadpole on his back.

Top: *Dendrobates fantasticus.* Photo by E. Zimmermann. **Bottom:** *Dendrobates reticulatus.* Photo by H. Zimmermann. Both these taxa are very close to *D. quinquevittatus.*

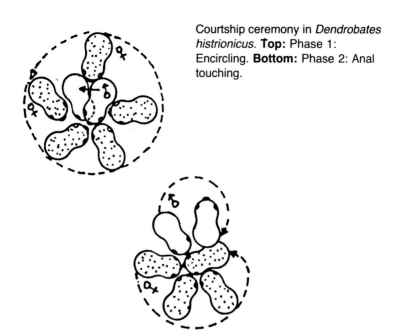

Courtship ceremony in *Dendrobates histrionicus.* **Top:** Phase 1: Encircling. **Bottom:** Phase 2: Anal touching.

inferior male that would frequently encounter a superior male within the confines of a terrarium should be separated. Otherwise it would remain in hiding while the rival male is present and so would not be able to feed properly. It would quickly lose condition and finally die. According to our observations, there are no aggressive actions between males and females or among several females. Females do not exhibit territorial behavior. There is no hierarchy among them, and they do not display any of the basic calls as described for males. However, if one male is kept together with two females in a terrarium of about 60 × 35 × 50 cm, predation on eggs and larvae is quite common. That is, if a female is ready to spawn and then finds eggs or larvae of another female already present at the selected site, these are simply eaten. A few hours later this female will then deposit her own eggs at the same site. Therefore, successful breeding requires that this species be kept in pairs.

The breeding season does not have any seasonal restriction; it extends throughout the entire year. However, breeding is only initiated at temperatures from 23° upward. Egg deposition is preceded by an extended courtship usually lasting over several days, during which both partners move around each other giving off a multitude of visual, tactile,

142

and acoustic signals. This serves to synchronize the breeding readiness of both partners. Several potential spawning sites are investigated during these courtship rituals. The progress of such courtship is clearly visible in the diagrammatic chain of typical reactions. This is followed by depositing the eggs and caring for the brood, events that were analyzed by us for the very first time.

Eggs are deposited during a 10- to 15-minute interval. The female presses her anal region firmly against the substrate, shakes her hind limbs, spreads the angled hind limbs slightly away from the body, and arches her back. After several body contractions the first few eggs are ejected; the hind limbs are then pulled in slightly and the animal slides forward a short distance in a circular motion around her own axis, either clockwise or counterclockwise. The process repeats itself until all eggs have been laid in several intervals. In this manner a female can produce up to 5 clutches per week, each consisting of 3 to 16 eggs (mean value 8 eggs; egg diameter 1.5 mm, gelatinous layer 5 mm). Once spawning has been completed, the female leaves the site. Then the male returns either immediately or within the next 3 hours for up to 30 minutes to water and fertilize the clutch.

In *D. histrionicus* all further brood care activities are taken over by the female, who keeps the clutch wet until the larvae hatch. However, in *D. pumilio* the male returns to the clutch to take care of it, and in *D. quinquevittatus* both male and female return at regular intervals to water the clutch until the larvae are beginning to hatch. Yet, in contrast to the dendrobatids in Stage 1, the clutch is not guarded by either male or female.

At the optimum temperature embryonic development extends over 10 to 11 days. Finally, the 11-mm larva bursts out of the gelatinous film with strong rudder-like movements of the tail and then lies out in the open. In contrast to the dendrobatids in Stages 1 and 2, the female now takes over the larval transport (the male in *D. quinquevittatus*). She crouches deeply into the group of larvae and, as soon as a larva has slipped onto the back of the female, she hops to a water-filled bromeliad stem. She moves backward toward the leaf base so that the larva can slide down and glide into the water. This procedure is repeated until each larva is transported individually to its own bromeliad leaf axil. However, brood care in

Top: *Dendrobates silverstoni*. Photo by E. Zimmermann. **Bottom:** *Dendrobates leucomelas*. Photo by H. Zimmermann.

Top: *Eleutherodactylus* sp., a barking frog. Photo by J. Dodd. **Bottom:** *Gastrotheca marsupiata*, the marsupial frog. Photo by H. Zimmermann.

dendrobatids of Stage 3 (that is, *D. quinquevittatus, D. reticulatus, D. fantasticus, D. pumilio, D. granuliferus, D. speciosus, D. histrionicus,* and *D. lehmanni*) is not restricted to spraying water over the clutch and transporting the larvae. Due to a well-developed memory, the female can find each larva again in its bromeliad stem and feed them at regular intervals with specially produced "food eggs." Therefore, the female of *D. histrionicus* can in this manner feed up to 4 larvae at various leaf sites by moving from one larva to another every other day in order to feed it with 3 to 7 "food eggs" (unfertilized eggs with reduced gelatinous covering).

The larva, which usually submerges to the deepest portion of the bromeliad stem when the plant is inadvertently touched or moved in some way, displays a characteristic behavioral change when another frog approaches. First the larva's body becomes motionless and rigid, and then suddenly it swims about in an uncoordinated fashion with quick tail movements, continuously thrusting the body partially above water level. This "begging behavior" by the larva was first described for *D. pumilio* by Weygoldt. While other frogs will then avoid the leaf funnel, the female arrow frog moves her head into the water several times and then turns around and sits with the posterior portion of her body above the funnel. The strong tactile stimulus caused by the larva attempting to move up against the female's posterior end (which could correspond to the tactile stimulus produced by the male during courtship) causes the female to eject eggs, in this case the "food eggs." This type of brood care (unique among vertebrate animals) lasts for about 11 weeks, at which time the 17-mm juvenile frog leaves its bromeliad stem for the very first time.

The reproductive behavior as well as the highly advanced brood care are very susceptible to disturbances (see the chapter on reproduction). When we started breeding these interesting frogs we had only one female of each variant, so we decided to remove each clutch from the terrarium in order to raise the eggs artificially for the purpose of establishing a larger breeding group. The initial difficulties encountered are described in detail in *Aquarien Magazin,* 1980 (#10). Essentially the procedure is as follows:

The clutch is removed from the terrarium and transferred into a small bowl that is filled with water to a level barely covering the gelatinous mass, leaving a small portion protruding. The container is covered with styrofoam or a plastic lid in order to achieve a humidity of nearly 100%. When kept at 18 to 24° C, the larvae will hatch after about 14 days. They are then transferred individually into small dishes of about 30 ml capacity. If the larvae from one clutch are kept together there is no intraspecific aggression. However, due to their metabolic wastes, which have a growth-inhibiting effect, the period of larval development for larvae kept in groups is about double that for larvae kept singly.

The daily artificial diet consists of a drop of egg yolk suspension each. According to our experience, any other food cannot be digested. The water has to be renewed every five hours after feeding. After 11 weeks the hind limbs can be seen in the larvae, which by then are about 21 mm long. The forelimbs appear after 15 weeks. At that point the markings on the back—black with yellow dots (*D. histrionicus*) or with red or yellow bands (*D. lehmanni*)—become more distinct. During the subsequent two weeks the tail will have completely regressed. At that time the water level is lowered from 3 cm to about 1 cm so the tiny (about 11 mm) juvenile frogs can climb up the side of the container and so avoid drowning.

It has been our experience that it is better to rear the young from that point on in small plastic containers of about 20 × 20 × 10 cm with little ventilation and a high relative humidity (at 22-25° C, 90-100% humidity). Since the skin of such small frogs is easily damaged, short, soft moss or a soft, damp cloth should be used as the bottom substrate. Also required are adequate hiding places in the form of wide but smooth leaves.

During the first 2 to 4 weeks following metamorphosis, it is particularly important that the juvenile frogs be given a multitude of small insects and mites. Therefore, they should be fed several times daily with springtails, tiny wax moth maggots, young leaf lice, and similar insects. Once the frogs have survived the first 4 to 5 weeks they will take the same food as their parents: *Drosophila*, newly hatched crickets, and similar items. Sexual maturity is reached after 12 months.

Left: *Hyla ebraccata*, the hourglass treefrog, with egg clutch. Photo by H. Zimmermann. **Bottom:** *Hyla arborea*, the European treefrog. Photo by G. Dibley.

Top: *Hymenochirus curtipes*, a dwarf clawed frog, pair in amplexus. Photo by R. Zukal. **Bottom:** *Hymenochirus* sp., a female dwarf clawed frog filled with eggs. Photo by Dr. H. Grier.

The following dendrobatid species (Stage 3) have a similar reproductive behavior with the same sort of brood care. (This was documented by us for the first time in detail for *D. quinquevittatus* and *D. reticulatus.*)

Dendrobates quinquevittatus
Maximum size: 1.9 cm.
Distribution: South America, entire Amazon region.
Habitat: Rain forests.
Egg development and larval size: 8–14 days; 1.1 cm.
Larval life and juvenile size: 1 ½ months; 1.1 cm.
Remarks: Tadpoles fed with food eggs by female, but also omnivorous; sexually mature at 8 months.

Dendrobates reticulatus
Maximum size: 1.6 cm.
Distribution: Peru.
Habitat: Rain forests.
Egg development and larval size: 8–14 days; 1.1 cm.
Larval life: 2 months; then see above.

Dendrobates pumilio
Maximum size: 2.4 cm.
Distribution: Central America.
Habitat: Lowland rain forests.
Egg development and larval size: 10–19 days; 0.6 cm.
Larval life and juvenile size: 2 months; 1.0 cm.
Remarks: Reared artificially in 3 months; tadpoles fed with food eggs by female. Artificial rearing possible only with egg yolk; sexually mature in 10 months.

Dendrobates leucomelas—Poison Arrow Frog
(Family Dendrobatidae)
Description: Size to 3.8 cm. Base color black with three yellow to orange crossbands that can also have black dots or irregular patterns. Hybridization with *D. truncatus* and *D. tinctorius* is possible.
Distribution and habitat: Venezuela and adjacent regions (Guyana, Brazil, Colombia). Damp lowland rain forests, altitude 50 to 800 m. Lives on the ground underneath wet rocks and on tree trunks and leaves.
Care: Food: Small insects, small worms, and spiders. Must be kept in a moist terrarium at temperatures from 23 to 28° C during the day, lowered to 18 to 20° C during the night.

Relative humidity 70 to 100% (spray with water twice daily). The substrate should be peat moss covered with green moss and have a section with leaves and a small water container. Plants can include *Scindapsus*, bromeliads, and similar species.

Behavior and breeding: Use a large, well–planted moist terrarium (70 x 60 x 70 cm) for breeding. This is a social species that is active during the day. We are already working with the third generation in our tanks after having started out with a single pair. Now, after 5 years, the first female is still laying eggs throughout the year (with interruptions). We were the first to record the reproductive behavior and the brood care strategy of this species.

At an age of about 15 months, the young males will start to vocalize for the first time with a croaking, staccato trill. This sound will develop during the following 2 to 3 months into the species–specific courtship call. Then a sort of hierarchy develops within a group of 2.8–cm females and their 3.8–cm mothers. One female will jump onto the back of another and attempt to hang on to its neck region and at the same time press the female underneath firmly to the ground. The pressure on the female underneath is further increased because the female on top will let herself fall onto the lower female using her entire weight by extending and bending the hind limbs. The lower female then continues to remain motionless for some time in the submissive position, the abdomen pressed firmly against the ground, even after the upper female has already jumped off.

The "fight" among males in this species is confined to a sort of competitive vocalizing. With the limbs spread and the body erected on extended forelimbs, the male—with grossly inflated black vocal sac—announces his mating willingness with a continuous trill (10 to 18 seconds, with about 30 pulses per second, main frequency 2–3 kHz at an air temperature of 25° C). This sound tends to vibrate the entire body of the frog. Other males in the same terrarium will then immediately interrupt their activities and take up calling positions about 20 cm away from the first male. From there they will then join in with a slightly varied pitch. A female ready to spawn will jump onto the back of one of the vocalizing males and move her upper body slightly up and down. This is followed by stroking over the male's back, alternately first

Top: *Hyperolius marmoratus*, the African marbled frog. Photo by B. Kahl. **Bottom:** *Hyperolius fusciventris*, an African reed frog. Photo by W. Mudrack.

Top: *Leptodactylus fallax*, a South American bullfrog, one of the barking frogs. Photo by Bonnetier and Lescure, courtesy Dr. D. Terver, Nancy Aquarium, France. **Bottom:** *Kaloula pulchra*, Malaysian narrow-mouthed toads. Photo by B. Kahl.

with one forelimb and then with the other. The male briefly shakes one leg and then the other and then continues to move on and repeats his call. The female follows him and again strokes his back. During such mating foreplay—an actual mating with copulation was not observed in any of the *Dendrobates* species described—which can last for several hours, the male moves throughout the terrarium in search of a suitable spawning site.

During this activity the pair is not necessarily left undisturbed by conspecifics. Frequently another female that is ready to breed follows the first female. She tries to jump on her back and attempts to push her head down against the ground. However, usually the first female cannot be suppressed during this situation. She crawls from underneath her rival and then continues to follow the (again) vocalizing male. On the way to the spawning site both partners frequently come face–to–face, nearly touching each other with their heads and skipping in a semi–circle around each other. Soon they will have found a suitable spawning site. In our terraria this is usually a petri dish covered by a coconut half–shell or some oak leaves. Once there, the male moves inside through the entrance hole and continues to vocalize until the female follows him inside. Frequently both partners move about inside this "cave," following each other in circles in a crouched position. Then the male reappears at the entrance and comes out. The female deposits 2 to 8 black eggs (about 3.5 mm in diameter and surrounded by a 10–mm gelatinous cover), and then also leaves the spawning cave.

The male picks up some water from the water dish and returns to the clutch a short time later in order to wet and fertilize the eggs (if we were to remove the eggs earlier, that is, right after they were laid, they would not develop). If one keeps only one pair in a terrarium of small to moderate size, it is possible to observe the brood care, provided the animals are not being disturbed. On the day of spawning the male will return to the clutch several times in order to keep the eggs wet. After that, however, for a period of about three weeks until the 18–mm black larvae have hatched, neither male nor female will return to guard the eggs. Apparently the large gelatinous cover provides sufficient protection against desiccation. After the larvae have burst out of the gelatinous cover and are resting freely, the male returns to the spawning cave. He positions himself right in the middle of the clutch and waits until one larva has slipped onto his back. This way, one–by–one, all larvae are transported on the back of the male to individual bromeliad stems or to small water holes in the ground.

Larval transport individually or in small groups is characteristic of all species of arrowfrog in Stage 2. The *Phyllobates* species usually carry several larvae at the same time, but *Dendrobates* species carry their larvae in most cases individually to small water holes. Presumably this is a protective mechanism against cannibalism, which occurs with high larval densities. Transporting the larvae to water concludes the brood care of this and the other species in this group.

If breeding is to be successful, the tadpoles, which are susceptible to diseases in dirty water, will have to be transferred into a plastic container. Rearing is essentially the same as described later for *Phyllobates tricolor*. However, the larvae of *D. leucomelas* and other representatives in Stage 2 are more carnivorous and will eat the eggs and larvae of other species as well as those from clutches of their own kind. With adequate food, sufficient hiding places, and lack of overcrowding (maximally 10 larvae in a plastic container 42 x 16 x 15 cm, water level 8 cm), the siblings are not cannibalistic. If after all these steps are taken there is still intraspecific aggression among siblings (as we have seen in the terraria of some of our acquaintances), the larvae have to be raised individually (see *D. histrionicus*). Fungus from left–over food spreads quickly to the larvae. Therefore, it is important that the water is changed completely every other day.

At a water temperature of 23 to 24° C during the day and 20 to 21° C, at night the tadpoles will grow to about 3.9 cm in 2 months. At that time the hind limbs begin to develop. After 3 months and a larval length of 4.2 cm, the forelimbs will break through and the yellow and black banded markings will become clearly visible. Within the following 14 days the tail is resorbed, gill breathing changes to lung respiration, and then the small (15 mm) juvenile frog climbs onto land. It should then be transferred to a moist terrarium, where it will soon pursue newly hatched crickets, leaf lice, and fruitflies. The adult markings are developed during the following 6 months.

Top: *Mantella cowani*, the variegated golden frog. Photo by H. Zimmermann.
Bottom: *Megophrys nasuta*, Asian horned frogs, pair in amplexus. Photo by H. Zimmermann.

Top: *Phyllobates boulengeri-espinosai* (complex), a poison arrow frog. Photo by H. Zimmermann. **Bottom:** *Osteopilus septentrionalis*, the Cuban treefrog. Photo by Ken Lucas at Steinhart Aquarium.

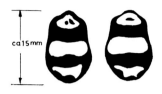

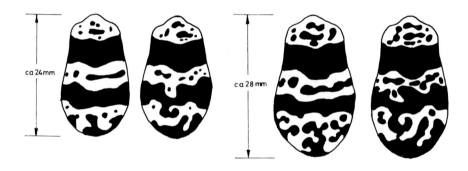

The dorsal markings of *Dendrobates leucomelas* change during the course of development. 15 mm—just after metamorphosis. 18 mm—after 3 weeks. 24 mm—after 3 months. 28 mm—after 6 months.

The reproductive behavior of the following species of Stage 2 is similar and they can be kept and bred under the same conditions. (We were the first herpetologists to observe and record the breeding behavior of *Phyllobates lugubris*.)

Dendrobates auratus
Maximum size: 4.2 cm.
Distribution: Central America.
Habitat: Damp forests.
Egg development and larval size: 11–14 days; 1.7 cm.
Larval life and juvenile size: 2 months; 1.3 cm. Sexual maturity in 1 ¼ years.

Dendrobates azureus
Maximum size: 4.5 cm.
Distribution: Surinam.
Habitat: Damp forests in savannah.
Egg development and larval size: 16–18 days; 2 cm.
Larval life and juvenile size: 2–3 months; 1.8 cm.

Dendrobates tinctorius
Maximum size: 5 cm.
Distribution: Guyana, northern Brazil.
Habitat: Rain forests.
Egg development and larval size: 14–21 days; 1.4 cm.
Larval life and juvenile size: 3–4 months; 1.7 cm. Sexual maturity in 1 ¾ years.

Phyllobates femoralis
Maximum size: 3.2 cm.
Distribution: South America, Amazon Basin.
Habitat: Lowland rain forests.
Egg development and larval size: 12-14 days; 1.1 cm. 1 ½ months; 1.2 cm. Sexual maturity in 10 months.
Larval life and juvenile size: 1½ months.

Phyllobates lugubris
Maximum size: 2.5 cm.
Distribution: Central America, Caribbean side.
Habitat: Lowland rain forests.
Egg development and larval size: 9–14 days; 1.1 cm.
Larval life and juvenile size: 2 months; 1.2 cm. Sexual maturity in 10 months.

Phyllobates terribilis
Maximum size: 4.7 cm.
Distribution: Western Colombia, Cordillieras.
Habitat: Rain forests.
Egg development and larval size: 11–12 days; 1.4 cm.
Larval life and juvenile size: 2 months; 1.5 cm. Sexual maturity in 1 ½ years.

Phyllobates vittatus
Maximum size: 2.9 cm.
Distribution: Central America, Pacific side.
Habitat: Lowland rain forests.

Left: *Phyllobates pictus*, a poison arrow frog, male guarding clutch. Photo by H. Zimmermann. **Bottom:** *Phyllobates pulchripectus*, a poison arrow frog. Photo by H. Zimmermann.

Phyllobates terribilis, a poison arrow frog. Photos by H. Zimmermann. **Right:** Female stroking male's back while he hunts for a suitable spawning site. **Bottom:** Female mating interference. One female is preventing another from mating by actually holding her down.

Egg development and larval size: 13–14 days; 1.2 cm.
Larval life and juvenile size: 1 ½ months; 1.3 cm. Sexual maturity in 10 months.

Eleutherodactylus johnstonei—Antillean False Arrow Frog

(Family Leptodactylidae)

Description: Size to 3 cm. Males somewhat smaller than females. Back brownish with variable dark markings; abdomen white to yellow.

Distribution and habitat: Lesser Antilles. Close to ground and in bushes.

Care: Food: Small insects such as *Drosophila,* crickets, and houseflies, as well as worms, water bugs, and snails. Should be kept in a moist terrarium with a small water section. Keep at temperatures from 23 to 25° C during the night. A good substrate is peat moss covered with living moss, oak leaves, and swamp roots. Plants can include bromeliads, begonias, philodendron, and similar plants.

Behavior and breeding: This primarily nocturnal species has developed a reproductive strategy that has made it totally independent of any open water. During the breeding season the males indicate their readiness to mate by vocalizing jointly with whistling sounds audible over a wide range. An approaching female is followed by a male to a spawning site, a damp, dark spot on the ground underneath leaves or under some tiles apparently selected by the female. The female deposits her eggs during the following 30 minutes or so (up to 30 eggs, 3 mm in diameter). They are fertilized by the male, which after that leaves the spawning site. The female remains with the eggs to provide brood care until the young hatch as juvenile frogs. Embryonic and larval development take place inside the egg until a fully developed frog bursts out of the egg membranes.

The eggs are yolk–rich. Development proceeds initially to a small, light brown larva; 9 days later the rudiments of forelimbs and hind limbs are already visible. These develop into functional limbs during the subsequent 4 days. The tiny frog, only about 3.5 mm long, bursts out of the egg membranes with the aid of an egg tooth on the upper jaw and slides out of the egg.

162

The young are then transferred to small rearing tanks (commercially available clear plastic refrigerator food containers have proven to be quite satisfactory for that purpose) that can be set up the same way as for adults. Initially the young frogs are fed springtails and small leaf lice; later this is followed by *Drosophila* and newly hatched crickets. With an adequate and diverse food supply the young will reach sexual maturity in 10 months.

As examples of convergent evolution to this leptodactylid from Central America, some African *Arthroleptis* species have also developed the same kind of reproductive biology. One example is *Arthroleptis wahlbergi*, which can be kept in semi–moist terraria at 22–26° C on peat moss covered with layers of leaves. The whitish eggs of this species (2.5 mm in diameter) are deposited in clutches of 11 to 30 in damp, decaying leaves on the ground. Two weeks later the rudimentary limbs are already visible on the developing larvae. Tail resorption starts 7 days later, and at the end of the fourth week after spawning the 6-mm froglets will burst out of their egg membranes. Rearing instructions are as described for *Eleutherodactylus johnstonei*.

Gastrotheca marsupiata (rhiobambae)—Marsupial Frog
(Family Hylidae)

Description: Size to 7 cm; males somewhat smaller than females. Dorsum light brown to green with brown to black longitudinal bands or patches. Abdomen yellowish white or patchy. Adhesive discs on extremities. Male with vocal sac and dark throat region. Female with a small horseshoe–shaped opening in lower third of back, the entry to the "pouch" that extends forward to the shoulder level.

Distribution and habitat: Central America, primarily Ecuador. Usually in damp forest regions.

Care: Food: Insects, worms, small pieces of beef. Should be kept in a moist terrarium at 20 to 25° C during the day and 15 to 25° C during the night. Relative humidity 70 to 100%. The substrate should be peat moss covered with green moss. A water container with water level at about 7 cm is necessary, and the frog also needs a climbing branch. Plants include *Scindapsus*, bromeliads, and others.

Behavior and breeding: The males of this species, which is ar-

Top: *Phyllobates vittatus*, courting pair. Photo by H. Zimmermann. **Bottom:** *Phyllobates tricolor*, male transporting tadpoles to water. Photo by H. Zimmermann.

Top: *Pipa pipa*, the Surinam toad. Photo by Ken Lucas at Steinhart Aquarium.
Bottom: *Phyllomedusa tomopterna*, the lemur treefrog. Photo by P. Weygoldt.

boreal and primarily nocturnal, signal their readiness to mate by a quick succession of clacking sounds followed by a prolonged croaking sound. If another frog approaches, the vocalizing one will attempt to grasp it. If this also happens to be a male, he will respond with a whistling protest sound, upon which he is then released again. If, on the other hand, it is a female ready to mate, the male will grasp her around the shoulder region. During the following amplexus the male on top of the female will turn his hands toward the inside so that a canal is formed. This canal is then filled with a whitish, milky liquid that is beaten into a foamy mass through alternate forward and backward movements of the hindlegs.

The female then arches her back downward so far that her cloaca is actually pointing upward. Then, at intervals of about 15 to 20 seconds, the female ejects 1 to 3 eggs per interval. These slide along the back toward the "pouch." With the toes of his hindlegs the male pulls open wider the entrance to the female's pouch and then actually pushes the eggs deep inside, first with one leg and then with the other. The eggs are fertilized during this process. The entire spawning act, which may produce up to 200 eggs (3 to 4 mm in size) lasts about 45 minutes.

This mechanism affords to the eggs excellent protection against detrimental environmental influences. During the following 4 to 6 weeks the eggs develop into tadpoles inside the female's pouch. (In the closely related *Gastrotheca ovifera* the eggs develop directly into juvenile frogs.) During the developmental period the skin on the back of the female becomes substantially stretched and small blisters become visible, each containing a tadpole. From that point on the female moves more frequently to the water section of the terrarium. Then one day she remains hanging on to the side of the water dish with the posterior portion of her body suspended in the water. With one toe of the hindfoot she partially opens the pouch on her back and so releases one tadpole after the other into the water. In our tanks we have observed up 100 tadpoles released by one female. Dead embryos and unfertilized eggs are removed at the very end, using the longest toe of one of the hind feet.

The tadpoles are about 1.8 cm long. They must be separated into several tanks equipped with filters and aerators. A suitable diet consists of a mixture of powdered tropical fish fry foods, algae, chopped tubifex, or small pieces of beef. In spite of stable environmental conditions, duration of larval development, as in many other frogs, is highly variable. Apparently the larvae can influence each other's growth through the excretion of metabolism–inhibiting products. At an average temperature of 23° C the first hind limb rudiments become visible after 22 days. About 45 days after spawning the front limbs break through, and metamorphosis is completed after a total of 49 days at the earliest. However, tadpoles that are inhibited in their normal development may require up to 110 additional days to complete metamorphosis into frogs. The young frogs (2.0 to 2.5 cm long) have juvenile markings in beige, greenish beige, green, brown or gold. They will then climb out of the water and onto floating cork islands.

The juveniles are then transferred into a terrarium with a moist foam rubber bottom and a few *Scindapsus* runners. Such a relatively sterile setup permits close monitoring of growth and development. Moreover, cleaning is greatly facilitated. At that stage the frog can already feed on houseflies and other insects of comparable size. With an adequate food supply there are no difficulties in raising these frogs. The more advanced specimens will reach sexual maturity in about 9 months.

Hyla arborea & H. meridionalis—European Treefrogs
(Family Hylidae)

Description: Size to 5 cm. Dorsum green to brown; a blackish lateral stripe from tympanum to pelvic region (*Hyla meridionalis* lacks the lateral stripe). Abdomen white. Adhesive discs on fingers and toes. Male with vocal sac and yellow throat.

Distribution and habitat: Large areas of central and southern Europe and western Asia including the Caucasus and Urals. Found in overgrown damp areas. During the breeding season found in bushes and brush at the edge of standing waters.

Care: Food: Insects and worms. Should be kept in a semi–moist terrarium or an outdoor terrarium at 18 to 25° C during the day and 12 to 18° C at night. The substrate should consist of peat moss covered with fresh moss; supply climbing branches, ferns, and epiphytes. The terrarium should include about ¾ land section and ¼ water section with *Myriophyllum*. Over–wintering (middle of November to middle of February) recommended at about 10° C.

Platymantis vitiensis, a terrestrial dwarf ranid frog from Fiji that lays eggs in a nest in pandanus leaves. The eggs are shown above. Photos by B. Carlson.

Top: *Rana "esculenta,"* the European water or edible frog, a fertile hybrid of *R. lessonae* and *R. ridibunda*. Photo by G. Dibley. **Bottom:** *Rana nigromaculata.* Photo by Ken Lucas at Steinhart Aquarium.

Behavior and breeding: Treefrogs are nocturnal animals. During the day they tend to sleep hidden under the leaves of bushes. However, sometimes they also like a bit of exposure to the sun, so they press their limbs tightly against the body and remain out in the open sun for the entire day. Outdoor terraria are ideally suited for keeping and breeding this species, although an indoor terrarium can also be used, provided it is about 66 x 50 x 50 cm and stocked with only 2 males and a female.

The mating season begins during late spring following the winter rest period. Males assume positions among the vegetation above the water and start to emit their courtship calls (15 to 30 sounds per vocal sequence). Apparently the males stimulate each other to join in unison. As was observed in toads, the innate sound releasing mechanism is still fairly unselective in treefrogs. That means that a vocalizing male will jump on anything that moves in front of it and is of about its own size and will attempt to grasp at it. If the approaching animal happens to be a male, it emits a prolonged protest or liberation call that it may repeat several times. At the same time its entire body vibrates strongly. Following this, it is released immediately.

A female ready to mate that approaches a vocalizing male will be jumped on by the male and grasped around its shoulder region. The female does not emit any sound during this activity. If a rival male approaches while the pair is in amplexus, the male on top of the female takes on a defensive posture. It pulls its hind legs tightly to the body, emits croaking sounds, and then kicks the hind legs explosively in the direction of the opponent, kicking him away. If the pair remains undisturbed, the female with the male on her back searches for a spawning site in an area in the water. With slightly bent hind limbs she feels around for suitable water plants to which she will attach her spawn. She then arches her back inward and downward and the first eggs are expelled. The male, without releasing his grasp, slides downward so that his cloaca comes into contact with the eggs as they are extruded. With strong movements of the cloacal lips he moves over the eggs and expels sperm in rhythmic contractions lasting from 30 to 60 seconds each. This procedure is repeated until all eggs have been ejected; only then will the male release his grip around the female. The female leaves the water, while the male often remains on top of the water plants and so remains available for further matings.

The yellowish white eggs, which are deposited in several clumps of 150 to 300 eggs, can be transferred into unheated aquaria (water temperature about 20° C) that have some water plants and are equipped with filters and aerators. The larvae hatch from their jelly–covered eggs after about 5 days and then drop to the bottom. The next day they will hang on the sides and among the water plants. Subsequently they can be fed with algae and finely powdered tropical fish fry foods. The hind limbs become visible after about 18 days, and after 33 days the front limbs will break through. About 3 days later the first 1.5 cm to 2.0 cm green or brown juvenile frogs (which still have a tiny tail remnant attached) will leave the water to climb onto a cork or styrofoam island. They can then be transferred to a semi–moist terrarium for further rearing. The diet should then be made up of fruitflies, houseflies, and meadow plankton, as well as small crickets. Sexual maturity is attained at an age of 2 to 3 years.

The following hylids have been bred under the same conditions:

Hyla cinerea (6.3 cm, southeastern USA, forests and meadows); *Hyla raddiana* (6 cm, Argentina); *Hyla regilla* (5 cm, western USA, forests and meadows); *Osteopilus septentrionalis* (14 cm, Cuba and Florida, forests and land under cultivation). The latter species, the Cuban treefrog, should not be kept together with smaller species. The females lay clumps of up to 2000 eggs in water. The larvae hatch in 1 to 2 days and complete their metamorphosis after 1 ½ months at a water temperature of 28° C. With an adequate diet, the 1.2 to 1.5 cm juvenile frogs will grow into sexually mature frogs in a few months. The large size of adult frogs allows them to take many foods, including other frogs and newly born mice.

Hyla ebraccata—**Hourglass Treefrog, Bromeliad Treefrog**
(Family Hylidae)
Description: Size to 3.7 cm, males somewhat smaller. Dorsum white to yellow-brown with dark brown spots. Upper thighs yellow. Abdomen white. Adhesive discs on all fingers and

Top: *Rana jerboa.* Photo by Ken Lucas at Steinhart Aquarium. **Bottom:** A leopard frog of the *Rana pipiens* complex, probably true *R. pipiens*, the northern leopard frog. Photo by B. Kahl.

Top: *Scaphiopus hammondi*, the western spadefoot toad. Photo by J. K. Lang-hammer. **Bottom:** *Rhacophorus leucomystax*, the Malaysian flying frog. Photo by H. Zimmermann.

toes. Males with yellow throat region and a vocal sac.

Distribution and habitat: Central America (Mexico, Guatemala, Costa Rica, Panama). Tropical primary forests; on grasses, bushes, and trees.

Care: Food: Insects. Should be kept in moist terrarium at air temperatures of 23 to 27° C during the day and 18 to 22° C during the night; relative humidity 70 to 100%; should be sprayed with water at night. Bottom substrate should consist of damp peat moss covered with fresh moss, climbing branches, bromeliads, *Scindapsus*, *Philodendron*, and *Dieffenbachia*. The back wall of the terrarium can be covered with cork tiles or similar material. There should be a small water container.

Behavior and breeding: These interestingly marked frogs remain hidden in water-filled bromeliad stems or in crevices in damp tree bark during the day. Only after the onset of dusk and darkness do they become active—a high relative humidity and heat are prerequisites. Mating readiness is signalled by the males, located on top of bromeliad or *Scindapsus* leaves, through squeaking individual calls (frequency 2.3 to 3.4 kHz) emitted from a grossly protruding yellow vocal sac. Vocalizing may be done either individually or in unison through apparent mutual stimulation. This is often followed by mild courtship fights in which one male jumps onto another one and attempts to push it off a leaf. However, usually such an attack is brushed off with a kick from a hind limb accompanied by the emission of a protest call.

A female ready to spawn is grasped firmly around the shoulder region. With the male on her back, the female searches through the terrarium for a bromeliad stem (axil) with a high water level inside or any other plant with leaves suspended above water. In our facilities about 150 eggs with comparatively little gelatinous material are attached to the upper or lower side of a leaf. In nature, newly hatched larvae slide into the water below or into the water-filled bromeliad stems. However, in a terrarium there is a great danger of desiccation in spite of frequent spraying with water, so we can cut the leaf off and place it inside a plastic container with water, so that the eggs are partially covered by water.

After 4 to 5 days at a temperature of 24° C, the 1-mm yellow-brown eggs with 2-mm-thick gelatinous cover develop into 5-mm-long larvae ready to hatch. They are then transferred into planted aquaria (25° C water temperature) with adequate filtration and aeration. The tadpoles (in shape very reminiscent of guppies) can remain virtually motionless in mid-water by using their rapidly vibrating tails. They maintain an individual distance of several centimeters from each other. This may be an indication of cannibalistic tendencies, because they will feed on diseased or dead siblings. Not surprisingly, these tadpoles are largely carnivorous; they should be fed mainly with tubifex, daphnia, or small pieces of beef. Some dried fish foods are also taken.

It will take at least 8 weeks until the first of the 1.2-cm juvenile frogs appears. From their first day on they exhibit the attractive golden-brown markings of their parents. Once transferred to a moist terrarium, our juveniles got a diet of small fruitflies, small crickets, and, later on, larger insects. They reached sexual maturity in 11 months.

Hyla variabilis (3 cm, Ecuador, rain forests) was bred under identical conditions. Spawning occurred in water, and the eggs were attached in clumps to water plants. The 0.6-cm larvae hatched in 3 days. They metamorphosed after 2 months into 1.2-cm juvenile frogs.

Hymenochirus boettgeri—Dwarf Clawed Frog
(Family Pipidae)

Description: Size to 3.5 cm. Dorsum brown with black spots. Abdomen and flanks yellowish white, mostly spotted. Of the five toes on the hind feet, the three inner ones have claws. There are broad webs between the toes. Males have distinctly visible postaxillary glands directly behind the bases of the front legs.

Distribution and habitat: Western Africa. Small and large bodies of water in forest regions.

Care: Food: Worms, water insects, small crustaceans, small pieces of beef, and small fishes. Should be kept in a well-covered aquarium with filtration at 20 to 24° C water temperature. The substrate can consist of sand, submerged tree roots, and smooth stones. Plants should include *Elodea, Myriophyllum,* or *Ludwigia.* Short-term increases in water temperature allegedly stimulate reproduction.

Xenopus laevis, the African clawed frog. **Top:** Mating pair in amplexus. Photo by R. Zukal. **Bottom:** Breeding female showing lobes of skin about cloaca. Photo by G. Dibley.

Chelonoidis carbonaria, the red-legged tortoise. *Chelonoidis* was formerly considered a subgenus of *Geochelone*. Photo at top by Dr. M. Freiberg; that at bottom by H. Zimmermann.

Behavior and breeding: In captivity the reproductive season of this genuinely aquatic frog extends throughout the year. When a male is ready to mate he raises his body on all four legs and emits a sequence of short, rapid courtship calls; they are reminiscent of the ticking of a clock. Another frog passing by will immediately be grasped. If it is also a male, it then produces a humming protest sound. This acoustic signal together with an olfactory secretion from the postaxillary glands identify it as a male, and it is then released immediately.

Females willing to mate recognize a male by its fanning limb motions together with trembling toes. They reply to an approach by a male with a lurching forward motion. This is followed by a rather primitive form of amplexus: the male does not grasp the female around her shoulder region as the higher anurans do, but instead grasps the pelvic region of the female with both arms. The male presses his body repeatedly against her back. He appears to stimulate her to spawn by simultaneously stroking her head and back several times with one of his hind limbs. The pair in amplexus then moves toward the water's surface in a wide semi-circle. Once there, the female will eject, belly-up, the first batch of eggs, which are immediately fertilized by the male. After that both of them move back down to the bottom. This cycle of movements, which is characteristic of all pipid frogs, is repeated several times in succession until all the eggs (as many or more than 1000 per spawning) have been ejected. When the female extends her legs stiffly to the back, shakes her forelimbs, waves her feet, trembles all over, and attempts to shake off the male on her back, the male will release his grip.

Since the parents are heavy predators on their own eggs, they should be removed to another tank. Alternatively, the 2 to 3 mm in diameter eggs floating at the surface can cautiously be scooped out and moved into a separate aquarium equipped with a filter and aeration as well as some plants. The larvae hatch (at a size of about 3 mm) after 5 days at a water temperature of 25° C. During the first few days they will remain attached to water plants or the sides of the tank by means of their unpaired adhesive glands. Afterward, they begin to swim about freely; at that stage they will have to be given adequate food. In contrast to most other pipids, the tadpoles of this particular species are carnivorous and right from the onset will have to receive sufficient meat in their diet.

The initial food should be infusoria, then cyclops and brine shrimp; finally, the tadpoles can receive the same kind of food as given to their parents. The hind legs appear after about 20 days, and another 15 days later the forelimbs break through. Metamorphosis into 2-cm-long frogs takes place during the subsequent 16 days.

The following African pipid species exhibit a similar reproductive behavior and can be kept and bred under identical conditions:

Hymenochirus curtipes
Maximum size: 3 cm.
Distribution: Congo Basin.
Habitat: Standing waters.
Egg development and larval size: 1 ½ to 2 days; 0.3 cm.
Larval life and juvenile size: 2 months; 2 cm.

Xenopus laevis—African Clawed Frog
Maximum size: 11 cm.
Distribution: Africa south of the Sahara.
Habitat: Quiet waters.
Egg development and larval size: 2 days; 0.4 cm; larvae are filter-feeders.
Larval life and juvenile size: 2 months; 1.3 cm. Sexually mature in 2 years. The tentacle-equipped larvae should be fed with paste fry foods or infusoria.

Hyperolius marmoratus—Marbled Frog
(Family Hyperoliidae)
Description: Size to 2.9 cm. Dorsal markings highly variable, usually whitish beige to dark green with black, dark brown, or red bands, stripes, or spots. Sometimes with yellow lateral bands. Abdomen white to yellow with or without black or red spots. Upper thigh as well as palms and soles red. Fingers and toes with adhesive pads. Male with protrusible, bright-yellow vocal sac. With more than 170 *Hyperolius* species, the individual species and forms are sometimes difficult or virtually impossible to distinguish. According to Schiotz (1975), the marbled frog belongs to the *Hyperolius viridiflavus* complex.

Top: *Chelydra serpentina*, common snapping turtle. Photo by Dr. M. Freiberg.
Bottom: *Chinemys reevesi*, Reeves's turtle.

Top: *Chrysemys picta*, the painted turtle. Photo by Dr. Herbert R. Axelrod. **Bottom:** *Chrysemys (Pseudemys) scripta elegans*, the red-eared slider or red-eared turtle. Photo by B. Kahl.

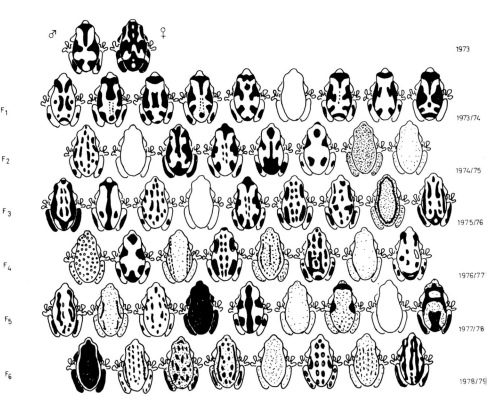

Variability in dorsal markings among the progeny of *Hyperolius marmoratus* over six captive-bred generations.

Distribution and habitat: Eastern and southern Africa. Overgrown swamps, ponds, and lakes with extensive reedgrass beds.

Care: Food: Insects such as mosquitos, flies, and crickets. Should be kept in a semi-moist terrarium at an air temperature of 23 to 28° C during the day and 18 to 28° C during the night. Spray with water at night. A small water section with *Myriophyllum, Eichhornia,* or similar plants is required. The land section can have either a foam rubber bottom for a sterile look or peat moss covered with fresh moss as a substrate. Plants can include *Scindapsus, Anubias,* or similar species. Climbing branches are readily used.

Behavior and breeding: During the day this species tends to sit on leaves, branches, or the container walls with legs and arms pulled so tightly against the body that the dark red

182

palms and soles are no longer visible. Once dusk sets in the animals become active again; they start hunting for prey and move actively about in the terrarium.

Breeding of our first 4 *H. marmoratus* generations always occurred during the winter months of November to February. However, since then—we now have generation F10—mating has gone out of synchronization, and the animals will now breed at any time of the year for a 2- to 3-month period. At that time males as well as females exhibit particularly attractive and intensely colored markings. The males take up their calling positions at the edge of the water container or on leaves and branches close to the water. They vocalize by emitting some piercing chirping sounds (about 1/10 sec long with a frequency range from 7 to 9 kHz) in rapid succession from a strongly protruded vocal sac. Once one male starts, others will usually join in. However, they will always attempt to sound off when another male has temporarily ceased calling, and they maintain a distance of at least 8 cm between themselves. If one male climbs onto the leaf already occupied by another vocalizing frog, that one changes his calling frequency and interval, begins emitting a territorial sound, jumps toward the intruder, and attempts to push him off. Such fights in conjunction with courtship vocalization may last for several days until a female is ready to spawn and approaches one of the males. The male then ceases his calls, jumps on the back of the female, and grasps her around her shoulder region. Then, with the male on her back, she hops to the water hole. In our terrarium this is a plastic container (14 x 14 x 8 cm) with a water level at 4 or 5 cm and containing a few strands of *Myriophyllum*. This is the spawning site where the female deposits on the average about 200 eggs (large females can produce up to 750 eggs) that have a diameter of about 1.5 mm. The eggs are laid in clumps on water plants or directly on the substrate. They are fertilized while spawning is still in progress, while the pair is still in amplexus.

After each spawning the water container is exchanged for a new one of identical size. The first tadpoles, still with light blue yolk sacs attached, will hatch after 3 to 4 days at an average water temperature of 24° C. They attach themselves to plants or the sides of the tank. Two days later they are transferred to a fish tank with water plants and equipped with aer-

Top: *Clemmys marmorata*, the western pond turtle. Photo by Ken Lucas at Steinhart Aquarium. **Bottom:** *Chrysemys (Pseudemys) scripta elegans*, the red-eared slider. Photo by M. Gilroy.

Top: *Gopherus polyphemus*, the gopher tortoise, hatchling. Photo by Dr. P. C. H. Pritchard. **Bottom:** *Geochelone elegans*, the Indian starred tortoise. Photo by B. Kahl.

ation and a filter. As the external gills begin to regress, the light brown tadpoles stay near the bottom among the water plants. The best diet consists of powdered dried tropical fish foods that are sprinkled on the surface and supplemented with chopped tubifex, scraped raw beef, and algae. Under crowded conditions the larvae tend to inhibit each other's growth.

About three weeks after hatching, when the larva has grown from 7 mm to 41 mm, the hind limbs appear. After another 11 to 12 days the front limbs break through, and the larva at that point has reached its maximum length of 46 mm. Two to 3 days later metamorphosis has been completed. The tail has been nearly resorbed, and the juvenile frog climbs onto floating cork islands or up the walls of the container. Larval development can be extended up to 101 days for some individuals instead of the normal 45 days. At that stage most of the 1.4-cm juvenile frogs still possess the nearly uniform golden-yellow coloration. They are then transferred into a semi-moist terrarium and are given a diet of fruitflies and meadow plankton; later on houseflies and small crickets are offered. Full adult color is not reached until after sexual maturity at an age of about 6 months.

The following hyperolids can be maintained and bred under identical conditions:

Afrixalus dorsalis
Maximum size: 3.5 cm.
Distribution: Western Africa.
Habitat: Savannah.
Egg development: 10 days.
Larval life and juvenile size: 2 ½ months; 1.2 cm.
Remarks: Oviposition on leaves over water; leaf is folded over clutch by male.

Hyperolius concolor
Maximum size: 4 cm.
Distribution: Western and central Africa.
Habitat: Gallery forests, savannah.
Egg development: 4 days.
Larval life and juvenile size: 2 months; 1.2 cm.

Remarks: Sexual maturity at 1 year. Oviposition on plants beneath and in water.

Hyperolius pusillus
Maximum size: 2 cm.
Distribution: Eastern and southern Africa.
Habitat: Plants at edge of water.
Egg development: 4 days.
Larval life and juvenile size: 1½ months; 1.0 cm.
Remarks: Oviposits on waterlily leaves or similar plants above water.

Hyperolius tuberilinguis
Maximum size: 3.4 cm.
Distribution: Southeastern to eastern Africa.
Habitat: Plants at edge of water.
Egg development: 5 days.
Larval life and juvenile size: 2 months; 1.4 cm.
Remarks: Oviposits on plants above water.

Hyperolius cinctiventris
Maximum size: 4 cm.
Distribution: Eastern Africa.
Habitat: Plants at edge of water.
Egg development: 8 days.
Larval life and juvenile size: 2 months; 1.2 cm. Sexual maturity at 1 year.

Kaloula pulchra—**Malaysian Narrow-mouthed Toad**
(Family Microhylidae)
Description: Size to 8 cm. Back dark brown. Head and lateral bands ochre-yellow. Shovel-shaped, sharp-edged digging tubercle (spade) on the heel. Males with a vocal sac on the throat studded with black granules. Fingers and toes with adhesive discs; webs very small.
Distribution and habitat: Southern Asia from the Indian subcontinent to southern China and the Indo-Malayian Archipelago. Tropical rain forests; during the breeding season found near small, standing waters.
Care: Food: Insects, spiders, earthworms. Should be kept in a moist terrarium at 24 to 28° C. The substrate can consist of a 20-cm layer of soil/peat moss mixture, oak tree bark, ferns, and climbing plants.

Top: *Terrapene ornata*, the western box turtle, hatchling. Photo by J. H. Mehrtens. **Bottom:** *Terrapene carolina*, the common or eastern box turtle, mating pair. Photo by B. Kahl.

Top: *Sternotherus odoratus*, the stinkpot. Photo by Dr. P. C. H. Pritchard. **Bottom:** *Testudo graeca*, the Mediterranean spur-thighed tortoise, with young. Photo by H. Zimmermann.

Behavior and breeding: These primarily ground-dwelling toads can dig into the ground with surprising speed with the aid of the digging "shovels" on the hind feet. This mechanism is similar to that found in *Pelobates, Scaphiopus,* and other spadefoots, and enables the toad to dig backward into the ground. When threatened these toads inflate themselves with air and give off a foul-smelling, slimy secretion. In their natural habitat they spawn during the rainy season; otherwise the dry soil would hold only temporary bodies of water. This situation can be simulated in a terrarium by keeping the animals dry initially and then transferring them to a terrarium with a large (half the size of the terrarium) water container with a water level of at least 6 cm and a water temperature of 26° C. *Scindapsus* vines and pieces of cork or bark can also be introduced to provide hiding places. While floating at the surface, the males begin to vocalize loudly in order to attract females. Spawning usually occurs during the following nights. The eggs are 0.7 to 1.4 mm in diameter within a thin gelatinous layer. They float at the surface and are strongly adhesive. They can be transferred—cautiously—to an aquarium with well-aged water.

At a water temperature of 20° C the larvae will become free-swimming in one day. They are fed with pulverized tropical fish foods that are sprinkled on the surface. The food is either sucked down from the water surface by the tadpoles or picked off the bottom once it has sunk. Larval development is completed in only 15 days, and the juvenile frogs (about 1 cm) climb out of the water and onto floating cork islands. Glass or plastic containers make suitable terraria for rearing the young frogs; all that is needed is a damp piece of foam rubber as substrate and some pieces of bark as hiding places. The container must be sprayed with water daily. Newly metamorphosed frogs should receive a diet consisting of small crickets, whiteworms, and leaf lice.

Kassina senegalensis—Running Frog, Senegal Kassina
(Family Hyperoliidae)
Description: Size to 43 mm. Dorsum gray with a dark median band and a dark dorsolateral row of spots on each side. Female slightly larger than male, during the breeding season with two pairs of leaf-like anal papillae. Male with vocal sac in the dark throat.

Distribution and habitat: Tropical Africa south of the Sahara. Savannahs and dry forests.

Care: Food: Insects, spiders, worms. Should be kept in a semi-moist terrarium at an air temperature of 20 to 28° C. Must be sprayed with water daily. A slightly lower temperature at night is permissible. The bottom substrate consists of a mixture of soil and peat moss covered with fresh moss and tufts of grass. A large, well-planted water section is required for breeding.

Behavior and breeding: Spawning conditions can be simulated in a terrarium by increasing the relative humidity through repeated daily sprayings with water. Then the males of this primarily nocturnal species gather around bodies of water, just as in the wild. Several males will form a vocalizing group. One male sets the tone by calling every 2 to 8 secs. Its calls stimulate replies from another male. In this manner distinctive vocalizing groups are formed, separated from each other by extended pauses. In essence then, these animals have the ability to adjust their individual calls (which are in the frequency range 1 to 2 kHz) in synchronization with those of their mates.

While vocalizing, a male raises up on its forelimbs and inflates its elongated body so that it appears to be nearly spherical. At the same time the vocal sac is vastly protruded. If a female approaches, the male will turn around toward her and encircle her, vocalizing throughout. Each sound is accompanied by the male's body moving excitedly forward and backward. Finally the male jumps onto the back of the female and grasps her around her pelvic region with both forelimbs.

The female will deposit up to 400 eggs in clumps on water plants. In order to do that it dives below the surface. The eggs are black on the upper surface, about 1.5 mm in diameter, and surrounded by a gelatinous layer about 2.5 mm thick. Small olive-brown tadpoles with large external gills will hatch in about 6 days. At that time they can be transferred to an aquarium equipped with a filter and aerator and well planted. After a few days the tadpoles can hold themselves nearly motionless in midwater by strongly vibrating the tips of their tails. They feed primarily on algae, water plants, and mosquito larvae; they are known to be aggressive predators! Metamorphosis is completed after about 3 months.

Top: *Trionyx spiniferus*, the spiny softshelled turtle. Photo by Dr. P. C. H. Pritchard. **Bottom:** *Testudo hermanni*, Hermann's tortoise. Photo by Dr. P. C. H. Pritchard.

> *Phrynobatrachus natalensis* (up to 3.6 cm; males smaller
> and with dark throat; savannah areas of Africa south of the Sa-
> hara) is maintained and bred under identical conditions. High
> humidity (simulated rainfall) releases breeding behavior. The fe-
> male oviposits in water (about 200-400 eggs 0.7 mm in size as
> film on water surface). Development takes 2 days; larvae 4-5 mm
> long. Tadpoles have to be put in separate tanks. Juvenile at
> metamorphosis 0.6 to 0.8 cm long.

Limnodynastes peronii—Australian Brown-striped Frog
(Family Myobatrachidae)

Description: Size to 6.5 cm. Dorsum light to grayish brown with irregular dark bands or spots. Abdomen white, frequently with brown spots. Males with grayish to yellowish throats and during the breeding season with stouter arms. Females with second fingers broadened.

Distribution and habitat: Eastern Australia, northern Queensland south to Melbourne. Shorelines of slowly flowing or standing bodies of water; frequently under rocks, rotting tree trunks, and rubble.

Care: Food: Insects and earthworms. Should be kept in an aqua-terrarium with decorative tree roots, rocks, or branches on the land section and *Elodea* and *Scindapsus* in the water. Acceptable temperatures 22 to 25° C, slightly lower during the winter.

Behavior and breeding: As with other frogs, the most interesting behavior can be observed primarily during the breeding season. This is a diurnal species. The males drift on the surface of the water or sit at the edge of the water and emit from their grossly protruded vocal sacs some metallic sounding "tok-toc" calls. Usually they do not call at the same time, but alternately. Only when there are a great many frogs at a spawning site can it happen that up to three frogs are calling simultaneously. However, a specific individual spacial distance is maintained among the males. If this distance is transgressed by a male, the offended male will immediately turn toward the offender and emit drawn-out territorial calls. If the transgressor still does not move off, the resident male will jump on him, grab him, and pull him below the surface. The following wrestling match in the water—during which fighting calls are given off by both males—usually lasts until one of them weakens so he gives up and swims away. The winner may follow the defeated male for a short distance, and then he returns to his vocalizing site to resume calling.

If a female approaches she is jumped on by a male and grasped below the upper forearms. During amplexus the male gives off a slimy substance that is whipped into foam by the female through alternate paddle-like motions of her upper arms. Into this foam, which usually floats at the surface and is anchored on water plants, the female will deposit up to 600 eggs. At a water temperature of about 23° C the tadpoles will begin to hatch after 3 to 4 days. Some powdered dried tropical fish foods are quite suitable as first foods. Within 3 months the tadpole develops into an approximately 1.5-cm juvenile frog that climbs onto land. From then on the main diet consists of fruitflies, houseflies, small crickets, and similar items.

Limnodynastes tasmaniensis (6 cm; female with thickened first and second fingers; Tasmania, Australia, swampy regions) can be bred under identical conditions. The foam nest at the surface contains up to 200 eggs (diameter 1.5 mm). The larvae hatch after 2 days; metamorphosis is complete after 3 months.

Mantella cowani—Variegated Golden Frog
(Family Ranidae)
Description: Size to 2.5 cm. Males somewhat smaller. Dorsum black with green spots at base of forelimbs and hind limbs. Green band from tip of snout through eye to base of forelimbs. Dorsal side of hind limbs golden, marbled with orange-black. Abdomen golden, marbled with greenish black. Male frequently with blue throat.
Distribution and habitat: Madagascar, Betsileo, Reunion. Tropical rain forest, on the ground among fallen leaves.
Care: Food: Insects, spiders, worms. Should be kept in a moist terrarium at an air temperature of 20 to 25° C during the day, with a slight temperature reduction during the night. The substrate (about 50% of entire area available) should have layers of damp peat moss "bricks" to form caves and other hiding places. The back wall can consist of cork or bark pieces with epiphytes. Plants may include *Scindapsus*

and similar species. The water section should have a water level of 4 cm and be planted with *Anubias, Cryptocoryne,* or *Myriophyllum.* The terrarium should be sprayed with water daily.

Behavior and breeding: This must be one of the most colorful ranids coming from Madagascar. It is primarily nocturnal and has been bred successfully in captivity. A moist terrarium of 50 x 50 x 50 cm is suitable for a breeding group of four males and a female. The male signals its mating readiness through repeated chirping calls, while the body is raised on extended forelimbs and the blue vocal sac is everted, balloon-like. If another male vocalizes close by, the first one will hop toward the rival and attempt to throw him on his back through a frontal assault. This is usually followed by a 6-minute-long wrestling match, a mutual assessment of each other's strength.

If a female approaches one of the vocalizing males, he will then hop toward her. If she moves on, the male follows her, making a circle around the female and stroking her body alternately with his forelimbs. Once the two animals reach one of the caves—they must be dark and in close proximity to water—the female will commence spawning within about 30 minutes. She expels 20 to 30 whitish eggs (diameter about 1.2 mm) that are presumably fertilized immediately by the male. The clutch of eggs should be left in the terrarium until the 6 mm tadpoles hatch in about 10 to 13 days. The larvae will slide into the water and can then be transferred in shallow plastic dishes with fresh moss or similar plants.

With a sufficient diet of fresh moss and vegetable-based dry fish foods, the young larvae will grow rapidly so that within 14 days after hatching the rudiments of the hind legs will be visible. About 11 days later the forelimbs will also break through. Now the water level has to be lowered, so the shallow dish is placed at an angle. After another 6 days the tail is resorbed and metamorphosis is completed. The tiny frogs (they are only 5 to 6 mm long) can then be transferred into a moist terrarium with peat moss layers, fresh moss, roots, and dead leaves. They are still brownish without green spots.

The initial diet should consist of a variety of small insects such as springtails, small leaf lice, and similar items; later *Drosophila* can be given. Coloration begins to change slowly after about 3 weeks. The first green spots become visible at the bases of the forelimbs; after 5 to 6 weeks they also begin to appear at the bases of the hind limbs. The upper thighs turn light brown and the body becomes black. Sexual maturity is reached in about 1 year.

The same breeding prerequisites prevail for the closely related *Mantella aurantiaca,* the **Golden Frog** (2.3 cm, males somewhat smaller; Madagascar; swampy forests). One clutch consists of about 20 to 60 whitish eggs with a diameter of about 1.5 mm. The tadpoles hatch after about 14 days at 21° C. These change into 9 to 11 mm brown juvenile frogs in about 7 weeks. Sexual maturity is reached after 1 year.

Megophrys nasuta—Asian Horned Frog
(Family Pelobatidae)

Description: Females to 16 cm, males to only about 9 cm. Dorsum light to dark brown; two distinct ridges on back; pointed skin projections ("horns") from nose and above eyes.

Distribution and habitat: Thailand, Malayan Peninsula, Indonesia, Borneo, Philippines. Cool, damp forests.

Care: Food: Insects, slugs, worms, small mammals. Must be kept in a moist terrarium at 18 to 23° C and a relative humidity of 80 to 100%. One half of the available area is to be used as a land section, with foam rubber on damp soil as the substrate; the other half is a water section with a water level of about 8 cm. Several pieces of cork bark and hardy plants such as *Scindapsus* or *Philodendron* are added. The terrarium should be sprayed with water at night.

Behavior and breeding: This species is primarily nocturnal. The males signal their readiness to mate by emitting short staccato-like "cock" calls. Just as in the other lower anurans, the female is grasped around her pelvic region by the male. Spawning usually occurs during the night following courtship. It is important to provide some sort of cave or archway made of pieces of cork or bark just above the water, because the 2 mm eggs (with only thin gelatinous envelopes) are attached in clumps to the damp roof of such a cave.

After about 11 days the young tadpoles come out of the gelatinous mass and slide down into the water on threads up to 15 cm long. At that stage the young should be transferred into an aquarium with adequate aeration and a filtration unit. Initially the 15 mm larvae will lie on the bottom, but after about 2 days the first swimming attempts are made; the larvae remain shy, hiding at the slightest disturbance. The characteristic funnel-shaped mouth of the tadpole will develop after another 8 days, and then the 20-mm-long brownish larvae will swim up to the surface. They are then ready to take their first food in the form of pulverized dry tropical fish food that is sprinkled onto the surface. This food is "sucked up" by contractions of the funnel-shaped mouth pointing at small food particles floating at the surface. When not feeding, the tadpoles tend to gather in dense formations along the floating cork islands.

At a water temperature of 24 to 26° C the light beige dorsal markings can be recognized after about 36 days. The hind limbs are fully developed after 75 days, and the 4.8-cm larvae begin changing their swimming behavior. Now they are more active, but they also tend to remain frequently at the bottom. The funnel-shaped mouth begins to regress at that point, and a few days later the forelimbs will break through. Now the water level has to be lowered to about 3 cm so that the 1.2 to 1.6 cm juvenile frogs can more easily climb onto land.

At this stage the young frogs can be moved into small plastic refrigerator containers (19 x 19 x 8.5 cm) with damp foam rubber as a bottom substrate. A small water dish should be added for bathing. At an air temperature of 15 to 22° C and a relative humidity of 70 to 90%, combined with a varied diet of crickets, wax moth larvae, and similar items, the characteristic horns on the nose and eyes will develop within 3 weeks. Sexual maturity is reached after 11 months at a size of about 8 to 8.5 cm.

Pelobates fuscus—**Garlic Toad, European Spadefoot**
(Family Pelobatidae)
Description: Size to 8 cm, males somewhat smaller. Dorsum colored variably; gray to whitish with dark or partially greenish spots and bands, frequently with red dots; abdomen white to yellowish. Males with oval arm glands during the breeding season.
Distribution and habitat: Western, central, and eastern Europe. Lowlands, in self-dug burrows and pits on sandy soil and in water holes during the mating season.
Care: Food: Insects, spiders, worms, and slugs. Should be

197

kept in a semi-moist terrarium (possibly outdoors) at 10 to 18° C during the night and 18 to 30° C during the day. The substrate should consist of sand and smooth stones and be at least 10 cm deep and moderately moist, neither wet nor dry. Provide a small water section. The toads should be given winter rest for 2 to 3 months.

Behavior and breeding: During the early morning hours this toad digs a burrow into sandy soil with the aid of its sharp-edged digging spades or tubercles. It remains in this burrow (often in asparagus fields) during the day and emerges at night to hunt for prey. If these toads are attacked they will inflate their body and give off a slimy, garlicky-smelling secretion from their skin glands. At the same time they will try to escape by kicking and scratching with their hind limbs.

During the mating season in spring, the males call for females at the edge of the water, emitting muffled "wock-wock" sounds in rapid succession. If a female approaches she will be grasped around her pelvic region by the male. Brownish black eggs about 1 mm in diameter are laid around water plants as egg strands 15 to 50 cm long and about 1.5 cm thick. About 5 to 6 days later the 4-mm-long larvae will hatch. They can be transferred to an aquarium equipped with adequate aeration, filtration, and plants. The initial food should consist of powdered dry tropical fish food, chopped tubifex, and tiny pieces of chopped beef heart. Metamorphosis will be completed in about 3 months, and the by then 3-cm juvenile frogs will climb onto land.

Phrynomerus bifasciatus—Red-banded Crevice Creeper, Two-banded Snake-neck Frog

(Family Phrynomeridae)

Description: Size of female up to 8 cm, male somewhat smaller. Dorsum black to brown with one red to yellowish longitudinal band along each side of body. Ventral side black. Limbs marbled red and black. Toes with small adhesive discs. Has a distinctively elongated body with movable neck. Males have black throats. Sometimes poisons with skin secretions other species kept in the same terrarium.

Distribution and habitat: Central, western, and southern Africa. Open brushland, usually in crevices underneath fallen tree trunks or in termite mounds, sometimes in low vegetation.

Care: Food: Insects. Must be kept in a semi-moist terrarium at air temperatures from 20 to 25° C, with a slight temperature reduction at night. Use a substrate mixture of peat moss and sand and provide a small water section.

Behavior and breeding: This species, like the spadefoots, burrows into soft ground, hind feet first. Deep below the surface in moist surroundings, this frog can survive even prolonged droughts. After heavy rainfall (simulated in a terrarium by repeated daily sprayings of water following a prolonged dry period) the males gather around water holes. There they emit long (about 2 seconds) trill sounds from their grossly protruded vocal sacs in order to signal their mating readiness. Frequently the males will call in choruses so their courtship sounds can be heard over long distances. This attracts those females within hearing distance.

The male grasps the female around her pelvic region during amplexus. Spawning occurs at the surface or among water plants protruding above the surface. About 1500 eggs are laid by one female. The eggs are light brown above, about 1.4 mm in diameter, and are embedded inside a 6-mm diameter gelatinous capsule. The first tadpoles hatch after about 4 days. They have external gills and an adhesive mouth. Much as in the larvae of *Hyla ebraccata,* they can—following resorption of the yolk sac—remain motionless in the water through constant rapid tail-beating. To feed, the larvae suck in algae and microorganisms. Metamorphosis is completed after about one month, and the 12 to 14 mm silver-gray and black banded frogs leave the water to live on land.

Phyllobates tricolor—Black and Gold Poison Arrow Frog
(Family Dendrobatidae)

Description: Size to 2.7 cm, females sometimes slightly larger than males and with wider median dorsal and lateral bands. Dorsum black or brown to brick red with white, yellow, or green median and lateral bands. Abdominal region and extremities yellow to white, frequently with a black or brown marbled pattern.

Distribution: Southwestern Ecuador, Pacific side of the Andes Mountains; in the proximity of water in areas with tall grass and pepper trees, 1200 to 1880 meters.

Care: Food: Insects such as *Drosophila,* small houseflies, small crickets, wax moth larvae, flour moth larvae, and similar items. Must be kept in moist terrarium at 22 to 30° C during the day and 18 to 22° C at night; 70 to 100% relative humidity. Spray twice daily with water. Plants may include ferns, bromeliads, *Ficus, Scindapsus,* orchids, and the like. Half coconut shells turned over and placed over petri dishes or layers of dead leaves make suitable spawning places. A water dish should be supplied.

Behavior and breeding: This poison arrow frog is active during the day, and it is one of the most adaptive and easily bred dendrobatid species. We have kept this species together with all of the smaller *Dendrobates* and *Phyllobates* species described in this book, even with *Phyllobates terribilis.*

Our breeding groups consist of from one male and one female to 6 males and 3 females accommodated in 70 x 60 x 70 cm to 120 x 50 x 110 cm terraria. In enclosures of this size one can easily and conveniently observe the interesting social and brood care behavior (Stage I, see the chapter on reproduction). Soon after introduction into a diversely decorated terrarium with several levels of terraces, sexually mature males will quickly establish territories. These consist of their sleeping place and potential spawning sites. They will move out of their respective territories only when in search of food or to court a female. If a male moves too deeply into another male's territory, the intruder will be warned acoustically by the offended male. Should the intruder refuse to move away, the defending male jumps toward him and onto his back, grasps him around the head, and pushes him to the ground or simply flicks him away with a forceful push with his head.

Females are not territorial; they remain primarily on the ground in the proximity of feeding places. Only at night will they move into a particular sleeping site on a bromeliad or other plant. In contrast to males, females do not vocalize, nor do they engage in any aggressive interaction with jumping toward conspecifics, head grasping, pressing down, or pushing except possibly over prey. If a female is ready to spawn, which can be as frequently as every 7 days, it leaves the ground and wanders through the territories of the males. At every territorial border the vocalizing resident male approaches the female. He tries to impress the female by raising his body on his forelimbs and stepping alternately from one foot to the other. When the female comes closer the male emits advertisement calls, turns around, and hops to a spawning site. If the partner and the spawning site are both acceptable to the female—which is not always the case—she will then follow the male. Once at the spawning site, which can be on a bromeliad leaf or inside a cave (such as a coconut half-shell turned upside down) the female either pushes herself underneath the male or the latter jumps on her back and grasps her around her head region. This so-called head amplexus is characteristic for all dendrobatids in Stage I. During this maneuver the male emits 2 or 3 long courtship calls. During the next 1 to 1½ hours he massages her body with his limbs, which apparently stimulates her to eject the eggs. By turning around her own axis and with alternately kicking legs and rhythmic body contractions she produces up to 43 eggs—over several intervals—that are immediately fertilized by the male. While the female remains with the clutch for

Territory formation in *Phyllobates tricolor*. Within the confines of a large terrarium the males (A-F) will defend certain areas containing potential spawning sites. Females (I-III) are non-territorial and prefer to remain close to the ground.

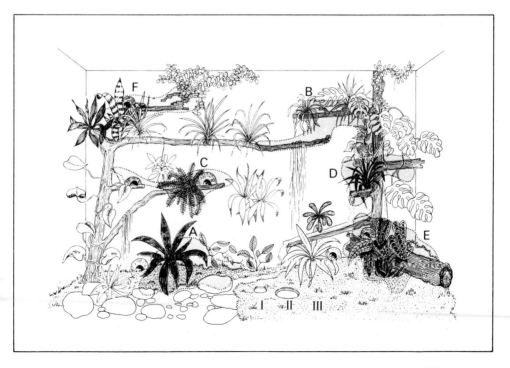

another 15 minutes or so after spawning, the male hops to the closest water hole (this could, for instance, be a water-filled bromeliad stem), where it picks up water for about 10 minutes and then returns to the clutch that by then has been abandoned by the female.

With his snout he pokes into the gelatinous mass and then finally settles down right in the middle of the clutch to spray water over the eggs at intervals. This procedure may be repeated several times on the day of spawning. As with all dendrobatid species of Stage I, the male takes over the brood care from that point on. He will remain close by until the larvae actually hatch (12 to 14 days at 24° C) in order to spray the eggs at regular intervals and to guard them. All other approaching frogs are kept away with head butting, jumping, head grasping, and pressing the opponent down while emitting territorial calls. However, despite all these brood care efforts, the males do not forget to feed and they also continue to court other females. Thus, it can happen that a male has to care for three separate clutches of eggs that are in the same general area. Once the larvae have hatched the male will provide their transport. He moves right into the middle of the clutch, and the larvae, responding to tactile and chemical stimuli, slide onto his back and adhere there. We have counted up to 30 closely packed larvae on the back of a single male. With very large clutches the male will have to make more than one trip. During the following two days the larvae are transported to larger water holes. The males move into the water, and due to the contact with water the larvae will gradually release their suction and start swimming away, straight into hiding. With this, brood care activities in this species are complete.

For further rearing, the 9 mm tadpoles (which as is common in Dendrobatidae have already developed a spiracle) can either be left in the water section of the terrarium or be transferred into special rearing containers. The latter guarantee maximum rearing results. Additional filtration and regular water changes are mandatory for the larvae to reach metamorphosis. Therefore, until metamorphosis, we keep the larvae in well-planted plastic containers at an initial water level of 1 cm that is then eventually raised to about 8 cm. Oak leaves are also given to provide additional hiding places.

The nonaggressive larvae should be fed a diet of algae, nettle powder, small pieces of beef heart, chopped tubifex, and various powdered dry tropical fish foods. Depending upon stocking density and the degree of pollution, the water should be changed once or twice a week. At an average water temperature of 24° C the larvae will grow to about 33 mm in 6 to 7 weeks. At that time the forelimbs will break free. Although the larvae at that time still will have a tail 21 mm long, precautions have to be taken that they do not drown. For that purpose they are transferred into plastic bowls with a wet fibrous bottom substrate and tilted at an angle. Once the tail has been completely resorbed the juvenile frogs are placed into plastic rearing terraria with moss or damp foam rubber as a bottom substrate and with bromeliads, fallen leaves, and pieces of cork bark to provide hiding places. A small dish of water should also be offered for bathing.

At that stage the young are already capable of eating insects in the size range of *Drosophila*. Their inconspicuous juvenile coloration of brownish green with two light lateral bands will change during the following 2 to 4 months to the more distinctive adult markings. Sexual maturity is reached about 9 months after metamorphosis. If there is a possibility that a particular clutch cannot be raised for some reason (not fertilized by the male, overcrowding, interference from other animals, etc.), attempts should be made for this and for any of the rarer species to raise the clutch artificially (for details see *Dendrobates histrionicus*).

The following dendrobatid species (Stage I) have similar reproductive behavior and can be kept and bred under identical conditions. (This was demonstrated by us for the first time for the following species: *Phyllobates anthonyi, Ph. boulengeri-espinosai,* and *Ph. pulchripectus.* According to Myers et al., 1978, the following 15 *Phyllobates* species should be included in the genus *Dendrobates: anthonyi, bassleri, bolivianus, boulengeri-espinosai, femoralis, ingeri, parvulus, pictus, petersi, pulchripectus, smaragdinus, tricolor, trivittatus, zaparo.*)

Colostethus inguinalis
Maximum size: 3.3 cm.
Distribution: Central and northern parts of South America.
Habitat: In the proximity of flowing waters.

Egg development and larval size: 10 to 14 days; 1.2 cm.
Larval life and juvenile size: 2 months; 1.3 cm.
Remarks: Provide rocky structures or clay pipes close to flowing waters for all *Colostethus* species; males *and* females territorial. Sexual maturity after 6 months. Brood care by female.

Colostethus trinitatis
Maximum size: 3.5 cm.
Distribution: Northern South America.
Habitat: Proximity of flowing waters.
Egg development and larval size: 13 to 18 days; 1.3 cm.
Remarks: Males and females territorial. Sexual maturity after 6 months.

Colostethus palmatus
Maximum size: 3.7 cm.
Distribution: Colombia.
Habitat: Proximity of flowing waters.
Egg development and larval size: 14 to 21 days; 1.3 cm.

Phyllobates anthonyi
Maximum size: 2.1 cm.
Distribution: SW Ecuador, NW Peru.
Habitat: Rain forests and adjacent areas.
Egg development and larval size: 10 to 14 days; 1 cm.
Larval life and juvenile size: 1½ months; 1.1 cm. Sexually mature in 12 months.

Phyllobates boulengeri-espinosai
(Complex of 2 almost indistinguishable species)
Maximum size: 2.1 cm.
Distribution: Western Ecuador, western Colombia.
Habitat: Rain forests.
Egg development and larval size: 16 days; 1.3 cm.
Larval life and juvenile size: 2½ months; 1.2 cm. Sexually mature in 12 months.

Phyllobates pictus
Maximum size: 3.1 cm.
Distribution: South America, Amazon Basin.
Habitat: Lowland rain forests.
Egg development and larval size: 16 days; 1.2 cm.
Larval life and juvenile size: 2 months; 0.8 cm. Sexually mature in 12 months.

Phyllobates pulchripectus
Maximum size: 2.7 cm.
Distribution: Brazil.
Habitat: Rain forests.
Egg development and larval size: 9 to 21 days; 1.1 cm.
Larval life and juvenile size: 2 months; 1.2 cm. Sexually mature in 12 months.

Phyllobates bassleri
Maximum size: 3.7 cm.
Distribution: Peru, Amazon Basin.
Habitat: Damp mountain forests.
Egg development: 14 to 17 days.
Larval life and juvenile size: 1½ months; 1.0 cm. Needs a very large terrarium.

Phyllobates trivittatus
Maximum size: 5 cm.
Distribution: Northern and central South America.
Habitat: Lowland rain forests.
Egg development: 14 to 19 days.
Larval life and juvenile size: 1½ months; 1.5 cm. Use a very large terrarium.

Dendrobates silverstoni
Maximum size: 4.3 cm.
Distribution: Peru.
Habitat: Damp evergreen mountain forests.
Egg development and larval size: 14 to 16 days; 1.3 cm.
Larval life and juvenile size: 1½ months; 1.5 cm. Use a very large terrarium.

Phyllomedusa tomopterna—Lemur Treefrog
(Family Hylidae)
Description: Size of male to 5.5 cm.; female to 6.5 cm. Dorsum green, flanks with yellow-brown bands. Upper arms, hands, and feet as well as inner areas of extremities and outside of hind limbs yellow brown. Abdomen yellow, throat white. Iris silvery gray. Webs between toes absent.
Distribution and habitat: South America, northern Amazon region. Tropical rain forest. Occurs among bushes, shrubs, and trees on leaves.
Care: Food: Insects such as flies, crickets, and wax moth larvae. Should be kept in a moist terrarium at 22 to 28° C and 40 to 50% relative humidity during the day (adequate ventila-

tion required), and 16 to 22° C and 70 to 100% relative humidity during the night. Provide a land section made up of peat moss covered with green moss and with plants with large leaves, such as *Philodendron, Dieffenbachia,* and *Scindapsus,* available. Needs a climbing branch. A small water section should be available, and the entire terrarium should be sprayed with water at night.

Behavior and breeding: These frogs sleep during the day, well-camouflaged underneath or on top of large leaves. In that position they pull their arms and legs so tightly against the sides of the body that none of the attractive markings can be seen. This sleeping position and a wax-like secretion from the skin protect against desiccation. They become active again at dusk, particularly after the terrarium has just been sprayed with water. At that time they climb about with slow and deliberate movements in search of food.

Breeding begins with the onset of the rainy season. This can be simulated in a terrarium by keeping the breeding animals (usually one pair is sufficient) under a continuous water spray or mist. In addition, the entire land section is replaced with a water section. Continuous rainfall can easily be created with a power filter, take up tube, and aquarium air stone. The air temperature should be maintained between 23 and 25° C. Under the influence of the high humidity thus created, the males alternately begin to give their courtship calls with the onset of dusk. If a frog of about identical size moves in the vicinity of one of those vocalizing males, one of the latter will attempt to jump on his back and try to grasp him. If the intruder also happens to be a male it will give off a protest call, upon which it is immediately released. However, a "silent" female ready to mate will be grasped around the shoulder region. The female may then wander around the terrarium for several days with the male on her back in search of a suitable oviposition site.

Eventually the clutch of up to 90 eggs is attached to a large leaf hanging over the water. The leaf is rolled up into a funnel shape over the eggs to protect them. The eggs are about 3 mm in diameter with a yellow-white coloration. They develop into larvae during the following 9 to 10 days. Then, after they have been sprayed, virtually all of them tend to burst out of their shells and slide down the leaf into the water below. At the time of hatching the yellow-brown larvae are

206

about 16 mm long. They should be split up into smaller rearing tanks planted with *Myriophyllum, Ceratophyllum, Elodea,* and other plants. The diet should consist of lettuce leaves (not chemically treated!), algae, nettle powder, and finely powdered dry fish foods.

The tadpoles maintain a characteristic oblique, nearly motionless position in the water with the aid of their rapidly undulating tail. They use this position to filter or siphon food from the surface. Their food demand is large, thus polluting the water, which then requires frequent changes. The juvenile frogs (2-2.5 cm) still have a substantial tail present but leave the water about 60 days later. They can then be placed in rearing terraria set up the same as for adults. Good ventilation is important! At first the young frogs have a tendency to hide until their tail is totally resorbed. Only then does it become necessary to provide a diverse diet of vitamin-dipped insects such as houseflies, crickets, moths, and similar foods. Initially the frogs still have their juvenile coloration (green back with whitish orange sides), but after 3 to 4 weeks the colors begin to change. The males begin to vocalize for the first time about one year after metamorphosis.

The following hylids can be bred under identical conditions:
Pachymedusa dacnicolor
Maximum size: 10.3 cm.
Distribution: Mexico.
Habitat: Rain forest.
Egg development: 7 to 9 days.
Larval life and juvenile size; remarks: 2 months; 2.2 cm. Eggs attached to leaf above water; adults will feed on young mice.

Phyllomedusa rhodei
Maximum size: 4 cm.
Distribution: Eastern Brazil.
Habitat: Rain forest.
Egg development: 14 days.
Larval life and juvenile size; remarks: 1½ months; 2 cm. Eggs attached to leaf above water, leaf folded over eggs by male.

Phrynohyas venulosa
Maximum size: 12 cm.
Distribution: Central America.

Habitat: Rain forest.
Egg development: 3 days.
Larval life and juvenile size; remarks: 1½ months; 1.5 cm.
Eggs as surface film on water; adults will feed on young
mice.

Smilisca phaeota
Maximum size: 8 cm.
Distribution: Central America.
Habitat: Rain forest.
Egg development: 2 days.
Larval life and juvenile size; remarks: 1 month; 1.5 cm. Eggs
as surface film on water; sexually mature in 1 year.

Pipa carvalhoi—Three-clawed Star-fingered Toad, Carvalho's Surinam Toad
(Family Pipidae)
Description: Size to 5.5 cm. Dorsum dark gray to brown with
black spots; abdomen white with dark patches. Three claws
on the hind foot. Female ready to spawn with ring-like cloacal swelling.
Distribution and habitat: Brazil. Small, nutrient-rich, exposed
ponds and lakes in an open or partially open landscape.
Care: Food: Tubifex, whiteworms, mosquito larvae, earthworms, small pieces of raw beef liver or beef heart, and fish.
Must be kept in an aquarium at 20 to 25° C equipped with a
filter. Use sand or gravel as the substrate. Hiding places such
as tree roots, robust plants, and rocky caves or grottos should
be provided.
Behavior and breeding: Although this species is active primarily during the day, visual perception is not well developed,
presumably because it lives mainly in very turbid waters. Instead, this frog has a well-defined olfactory sense and a lateral
line system as well as thermo- and mechanoreceptors on its
finger tips and along the snout.

The frog swims along the bottom with its forelimbs extended forward, stopping from time to time to make pushing
movements toward the mouth, testing with its olfactory re-
ceptors the water current. If the frog has perceived food, it
presses its head firmly against the bottom and through sudden snapping (vacuum effect) sucks in the food and swallows
it. Any sand grains or other debris taken in at the same time

are expelled again together with the water in the mouth.

This species has been bred successfully in captivity. We currently have the generation F4 in our tank. The aquarium has dimensions of 50 x 30 x 25 cm. Courtship vocalization starts about a day before spawning is to occur. With their forelimbs extended onto the bottom or resting on water plants, the males give off a courtship call that lasts for up to 16 seconds and consists of a series of short "ticks" reminiscent of winding up a clock. Any other frog swimming past is followed, touched with forelimbs and finger tips, and then grasped around the pelvic region. If this specimen then extends its legs rigidly toward the back, trembles, and attempts to rid itself of the male on its back, it is then either a female not yet ready to spawn or another male. If it is indeed a male, it will also emit a protest call, upon which it is immediately released again.

However, a female ready to spawn (with slightly swollen dorsal skin and a ring-like swelling around the cloaca) will be firmly grasped for the next few hours. At that point spawning commences, a procedure that is similar in all pipid frogs. While still in amplexus, the pair will swim a few circles vertically or at an angle through the entire aquarium. Shortly be-

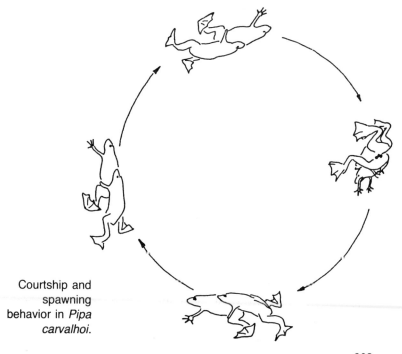

Courtship and spawning behavior in *Pipa carvalhoi*.

fore going down toward the bottom the male raises its body slightly off that of the female. At that point the female ejects a number of eggs, which are immediately pushed onto her back by the male and are fertilized at the same time. Subsequently the male distributes the eggs over the entire back of the female with stroking motions of the hind limbs.

During this ritual, which is repeated several times, up to 200 white eggs are produced. During the following days the eggs become embedded in the dorsal skin of the female. During this process the gelatinous envelope is pushed up and so forms a sort of lid above each egg chamber that falls off after 2 to 4 days. At that time the egg chamber has a nearly invisible opening to the outside. The chambers are only 0.4 mm wide. Embryonic development occurs inside the chambers, where the eggs are protected against detrimental environmental influences. About 2 weeks after they have become embedded, one can see many small "swellings" on the back of the female. The larvae have grown sufficiently to press the dorsal skin outward, and they also begin to move. A few days later the small skin mounds burst open and tadpoles 11-12 mm long slide out of the individual egg chambers and begin to swim in search of a hiding place. After the larvae have hatched, the back of the female has tiny craters that will quickly close over. One to two weeks later the female is ready to spawn again.

In order to breathe, the larva has to come to the surface, but then it sinks again immediately. The tadpoles should be removed without delay (they are prey for their parents) and transferred into a separate aquarium with water plants, a filter, and moderate aeration. About 3 to 4 days after hatching, the larvae swim about freely and filter-feed on pulverized tropical fish food, which is taken off the water surface. The larva's body remains transparent throughout the entire larval development, and blood vessels and internal organs are clearly visible. After 8 weeks at a water temperature of 24 to 26° C the hind limbs can be distinguished, and after another 2 weeks the forelimbs. At that stage the larva has a total length of 30 mm. The tail is resorbed during the following 10 days, and soon thereafter the young frogs (15 mm) begin to search for food such as small worms, daphnia, cyclops, and similar items. They will reach a size of 35 to 45 mm within 5 to 8 months; at that stage they are sexually mature.

> *Pipa pipa,* the Surinam Toad (20 cm; northern South America; standing or slow flowing waters) can be bred under largely the same conditions, but requires a much larger tank. In this species the entire embryonic and larval development takes place inside special egg chambers on the back of the female. The young toads will emerge after 80 to 140 days. Rearing as above.

Rana "esculenta"—European Water Frog
(Family Ranidae)

Description: Size to 12 cm. Dorsum green to brown with dark patches, frequently with light-colored median dorsal band; abdomen whitish, frequently with dark patches. Underneath upper thighs marbled with brown, black, and yellow. Males with two whitish vocal sacs and during breeding season with gray horny swelling on first finger.

Rana "esculenta" was once treated as a valid species, but now it is considered to be a complex hybrid form (Arnold & Burton), since research has shown (Berger, Heusser & Guenther) that viable *"esculenta"* specimens are obtained primarily from crosses of *R. "esculenta"* x *R. lessonae* or from *R. ridibunda* x *R. lessonae.* It is now possible to distinguish between these three "species" on the basis of biometric, serological, and genetic characteristics. Therefore, a breeding group should consist of a mixture of green frog populations, such as *Rana "esculenta"* and *Rana lessonae.* On the basis of more recent investigations it has been found that some genetically pure *R. "esculenta"* populations are also capable of reproducing. (*Rana "esculenta"* can be distinguished externally from the smaller pond frog, *Rana lessonae,* by its larger hind limbs and the smaller heel spur, ⅓ to ½ times longer than the first hind toe.)

Distribution and habitat: Europe, from Sweden to Sicily, and from Great Britain to western Russia. Standing and flowing waters in lowlands and highlands.

Care: Food: Spiders, worms, and insects such as flies, crickets, and mealworms. Should be kept in an aquarium or outdoor terrarium provided with floating cork island(s) or other floating objects and also *Elodea* and other water plants. Water temperature 13 to 21° C, air temperature 10 to 18° C at night and 18 to 30° C during the day. Should be over-wintered in a damp moss and leaf mixture at about 5° C.

Behavior and breeding: This species is active primarily during daylight hours. It is frequently found sunning itself along bodies of water. When approached it will initially remain motionless and so rely on its grayish green camouflage coloration. However, if a flight distance of about 1.5 m is exceeded it jumps into the water and tries to hide among plants or by digging into the bottom substrate.

Mating season for this species is in spring. The males will start to vocalize with laterally extruded vocal sacs, usually from locations around water holes. They may also float on the water in search of a female. During this period males are territorial; they may occupy a section along the shoreline and then defend this area against intruding males. If, however, a female approaches the male will swim toward her. Usually the female then tries to flee, but the male follows and grasps her around the shoulder region. A female ready to mate will soon thereafter start to spawn. Rather sizeable egg clumps are formed and at first float at the surface. They may contain up to 15,000 black eggs with a diameter of 1.5 mm, surrounded by gelatinous envelopes measuring about 6 mm in diameter. A few hours after spawning the eggs sink to the bottom.

At a water temperature of 20° C the black tadpoles (6-7 mm long) hatch after 5 to 8 days. Initially they adhere to the gelatinous mass, but eventually they fall to the bottom. Soon thereafter they start swimming freely. The water should be filtered and aerated and kept at a temperature of about 23° C. Water plants are recommended. A suitable diet includes algae, detritus, small pieces of raw meat, chopped tubifex, and various tropical fish foods. The hind limbs appear after about 85 days, and after 93 days the front limbs break through. Three days later the tail is completely absorbed and the 2-cm juvenile frogs hop ashore or onto floating islands in the aquarium. They can be raised there. According to their individual sizes they should be fed on insects, worms, or small pieces of raw meat (offered "in motion"). Sexual maturity is reached in 2 to 3 years.

*The **Rana pipiens** complex (12.8 cm; leopard frogs; most of North America, in the proximity of water) is bred in the USA for use in scientific research.*

REPTILES
Tortoises and turtles (Testudines)

Chelonoidis carbonaria [Geochelone]—**Red–legged Tortoise**
(Family Testudinidae)
Description: Size to 50 cm. Carapace dark brown to black with brownish yellow dots in the middle of the dorsal shield; plastron yellowish with black spots. Reddish to yellow scales distributed irregularly over the head and extremities. Male with indented plastron.
Distribution and habitat: South America east and west of the Andes. Tropical rain forest, on the ground.
Care: Food: Sweet fruits, vegetables, lettuce, commercially prepared dog food, fish, meat, freshly killed pink mice. Should be kept in a semi-moist terrarium with a radiator or sun light, the temperature about 26° C during the day and about 20° C at night. The substrate can consist of damp peat moss. Tree roots or logs should be provided, as should a water bowl. The terrarium should be sprayed with water in the morning.
Behavior and breeding: This species is active primarily during the day. It can really only be bred in large terraria. The breeding season extends throughout the year. During the initial courtship proceedings the male follows a female walking in front of him, sniffing at her hind legs and cloacal region and ramming his shell against her carapace. He also walks around her with his head and neck extended rigidly, moving them up and down or from one side to the other. During this behavior the male emits cackling sounds and bites the female on the head and front legs, forcing her to remain stationary. Then he mounts her from behind, resting his forelegs on her carapace and seeking cloacal contact with his tail. This is followed by the actual copulation, during which the male gives off a groaning sound with each push.

A few weeks later the female digs a pit about 13 to 22 cm deep with her hind legs and there deposits 6 to 7 eggs. Then the pit is carefully closed again. The eggs now have to be dug up again—cautiously—and their upper side clearly marked with a pencilled cross. They are then placed, correctly oriented, into a plastic box filled with damp peat moss and then transferred to a Type I brood container.

At 30° C and a relative humidity of 80%, the incubation period lasts for 106 to 184 days. At that stage the young tortoises will slit their egg shells open but will still remain inside for another 2 to 3 days until their yolk sacs have been completely absorbed. Only then will the young tortoises leave the shell. They are about 4.7 cm long and 4.4 cm wide, with a weight of 32 grams. Upon hatching they should be transferred to a semi-moist terrarium, where they can be given the same diet as the adults.

> *Kinixys belliana (to 20 cm; Africa south of the Sahara and Madagascar; semi-moist grassy bushland areas) can be bred under the same conditions. The incubation period lasts about 123 days at a temperature from 25 to 30° C. At the time of hatching the young are 3.6 to 4.7 cm long.*

Chinemys reevesi—Reeves's Turtle
(Family Emydidae)

Description: Size to 22 cm. Carapace dark brown to black. Plastron yellow with dark spots. Stout head; webs small; extremities and neck dark brown. Two to 3 yellow longitudinal lines on neck; throat spotted with yellow brown. Male with slightly indented plastron and enlarged base of tail.

Distribution and habitat: Central and southeastern China, Korea, and also introduced into Japan. Streams, ponds, canals, rice paddies.

Care: Food: Mosquito larvae, tubifex, fish, shrimp, mussels, snails, earthworms, small pieces of beef liver and heart. Should be kept in an aquarium with a water temperature of 20 to 27° C; can be kept outdoors during summer months. Requires a water area of not less than 1 sq. m. for breeding and a small land section with suitable places for depositing eggs (damp to moist sand). Can be given a rest period during winter months in damp wood shavings at 10° C and in darkness if possible.

Behavior and breeding: This species is active primarily during the day. It can easily be kept together with other species. However, it is possible that fighting will occur among members of the same sex, especially in smaller tanks where there are few opportunities to hide or escape. Most common is chasing and biting of a smaller or weaker animal. Such an animal should be removed in order to avoid stress problems. Juvenile males sometimes engage in ritualistic fighting. Two opponents approach each other with body raised on all four legs and head and neck extended rigidly forward. They run toward each other and with a final, lurching leap they bump head against head. At the same time the opponents open and close their jaws to the accompaniment of strong guttural movements and simultaneous "water chewing."

During the mating season a courting male approaches the female head-on with a stalking, slightly wobbling gait and finally bumps into the female nose to nose—in a manner similar to the ritualistic fighting among males—with the characteristic chewing motion of the jaws ("water chewing"). Females unwilling to mate will then attempt to chase the male away with biting, but those ready to breed tolerate such approaches and will copulate.

A few weeks after copulation the female begins to search for a damp area in the aquarium's land section that is to serve as the egg-laying site. Once this has been found, she will dig a pit about 5 to 10 cm deep and there deposit 2 to 3 eggs 32-33 mm long and 20-21 mm in diameter. These eggs should be carefully removed from the soil and the upper side should be marked with a pencil. (This is to avoid incorrect repositioning of the eggs during the subsequent incubation, which could lead to the death of the embryo.) The eggs are placed inside a Type I brood box, which offers better environmental conditions and provides for better monitoring and control than leaving them in the terrarium. There the eggs are put into a damp soil substrate at a temperature between 24 and 32° C. The young will hatch in about 79 days. The actual hatching procedure can last from 1 to 2 days. The young turtles are 28 mm long, 22 mm wide, and 16 mm high; their weight is only 6 grams. Flattening of the soft carapace—originally oriented lengthwise inside the egg shell—occurs during the next 1 to 3 days. For rearing details see *Chrysemys scripta.*

Chrysemys (Pseudemys) scripta—**Pond Slider**
(Family Emydidae)
Description: Size to 25 cm. Carapace olive-gray to brown, fre-

quently with dark vertical bars. Plastron yellow with black spots. Head, neck, and legs with yellow-green longitudinal stripes, in some subspecies with bright red or yellow lozenge or spot in stripe behind eye. Male with longer front claws and longer tail with a thicker base than female.

Distribution: Much of southeastern and central USA south to northeastern Mexico. Standing or slowly flowing water with heavy underwater vegetation.

Care: Food: Worms, water insects, snails, fish, small pieces of raw meat, water plants, lettuce, flowers, fresh vegetable shoots, commercially prepared dry turtle food. Should be kept in an aquarium and (during the summer months) in an outdoor terrarium with a large water section; temperature 25 to 30° C. Tree roots or similar objects can be added as basking places and to provide transition from the water to the land section. Provide a thick (25 cm) damp sand substrate in the land section. A heat source and light for basking should be given if unfiltered sunlight is not available.

Behavior and breeding: This aquatic turtle is primarily active during the day. It likes to haul out onto land or roots to soak up sunlight or the warmth under a terrarium radiator. The courtship details are similar for all species in this genus. The male swims toward the female from behind and sniffs at her cloaca. Then courtship commences with a frontal approach and trembling of the male's front feet and claws. Any future action is determined by tactile contact to head and throat. If the female is not yet ready to mate she will snap at the male, which moves off slightly to one side to offer his carapace to the female. After a courtship that may last from a few hours to several days, the male eventually succeeds in getting the female to pull in her head. Then he swims toward her from behind, climbs onto her carapace, and wraps his tail around hers; after cloacal contact, copulation takes place.

A few weeks later the female displays increasing interest in the land area of the terrarium as she keeps searching for a suitable site to deposit her eggs. If she must use an area that may not be damp enough, she empties her bladder onto it to get the right sand dampness to assure the correct incubation environment for the eggs. With alternate digging movements of the hind legs, using the upper side of the foot, the female then begins to excavate a pit. Once the pit has been completed the female ejects her eggs at intervals of 3 to 10 min-

216

utes, up to an average of 5 eggs (2-22) per clutch. As laid, each egg is caught in the webs of the hind feet and then slides into the pit. Once laying has been completed, the pit is closed again with alternate shovelling movements of the feet and smoothed over with rocking movements of the plastron. Afterward there is virtually no evidence of the nest, and the female has fulfilled her obligations. In order to be able to detect the nest again one must have actually observed the eggs being laid, then afterward the site must have been marked with a stick or small stone.

Later the eggs are carefully dug up again and on each the side facing up is marked with a pencil. The eggs are now embedded in damp peat moss inside a clear plastic box and placed inside a Type I brood container. Hatching occurs after about 93 days at a temperature of 25 to 30° C. To raise these newly hatched turtles they are transferred into aquaria with a small land section, a heater, and a filter. The diet should consist of daphnia, tubifex, water snails, fish, and commercially prepared dried turtle food. Regular UV radiation (prophylactic against rickets) and a varied diet are prerequisites for proper growth and development. Under these conditions sexual maturity will be attained in about 4 years.

The following aquatic emydid turtles can be kept and bred under identical conditions:

Chrysemys picta—Painted Turtle
Maximum size: 25 cm.
Distribution: Southern Canada and much of USA.
Habitat: Standing and slowly moving water with dense vegetation.
Clutch size: 7.
Incubation period: 68 to 80 days at 27° C; 2.5 cm.

Clemmys marmorata
Maximum size: 18 cm.
Distribution: Western USA.
Habitat: Waters with dense vegetation.
Clutch size: 11.
Incubation period: 73 to 80 days at 25 to 30° C; 2.5 cm.

Emydoidea blandingii—Blanding's Turtle
Maximum size: 25 cm.

Distribution: Southeastern Canada and northeastern-central USA.
Habitat: Waters with dense vegetation.
Clutch size: 8.
Incubation period: 80 days at 23° C; 3.7 cm.

Mauremys nigricans
Maximum size: 20 cm.
Distribution: China, southeastern Asia.
Habitat: Waters with dense vegetation.
Clutch size: 3.
Incubation period: 67 to 79 days at 23 to 30° C; 4 cm.

Mauremys caspica—Caspian Turtle
Maximum size: 25 cm.
Distribution: Balkan countries to the Middle East and south-western Asia.
Habitat: Waters with dense vegetation.
Clutch size: 10.
Incubation period: 78-81 days at 29 to 30° C; 2.5 cm.

Emys orbicularis—European Pond Turtle
(Family Emydidae)
Description: Size to 30 cm. Carapace black to dark brown with a pattern of lighter—usually yellow—dots or lines; head, neck, limbs, and tail marked similarly; plastron yellow with black spots. Plastron in males concave; tail of male longer than that of female.
Distribution and habitat: Europe (except northern Europe and some areas in central Europe), western Asia, northwestern Africa. Standing or slow-flowing waters with dense shore vegetation.
Care: Food: Worms, aquatic insects, fish, pieces of meat, freshly killed pink mice. Requires a large aquarium with a land section, preferably a garden pond or a large water section in an outdoor terrarium. Air temperature 20 to 28° C. Over-wintering in water or damp leaf-moss mixture at about 5° C is recommended.
Behavior and breeding: This species is active primarily during the day. It tends to haul out onto logs or driftwood or to come onshore during the early daylight hours in order to soak up sun to reach the optimum body temperature. Only then will this species return to the water to hunt for prey. If these turtles are disturbed while sunning, they simply let

themselves fall back into the water, dive, and then hide among water plants or bury themselves in mud at the bottom.

The breeding season extends from spring until fall. Mating starts off with some foreplay. The male approaches the female from behind, examines her cloacal region, swims around her, and then makes a frontal approach. He stops about 2 to 3 cm in front of her snout and then—sometimes briefly moving back—begins to court her with forwardly extended forelegs and slightly vibrating forefeet held at an angle. During this behavior there is frequent tactile contact with the throat and head region of the female.

If the female is ready to mate she will retract her head into the shell and remain stationary for awhile. The male then swims immediately over her carapace and grasps the edges of the shell. With his neck extended far forward he makes jerking motions, sometimes biting the neck of the female, and wraps his tail around that of the female; copulation then occurs.

About 4 to 6 weeks later the female begins to examine the land section closely in search of a suitable place to lay her eggs. When such a place has been found she will initially drill a hole with her tail into the damp substrate and empty water from her bladder into the hole. Then she will begin to excavate a hole about 8 cm deep and 14 cm wide with alternate shovelling movements of her hind legs. In the course of one to 3 hours the female will produce up to 16 eggs, one egg at a time. She catches the egg with her folded hind legs and then lets the eggs slip gently into the pit. Once egg-laying has been completed she covers the pit again rather neatly. This site should be properly marked so that the eggs, which are 31 to 39 mm long and 19 to 25 mm wide, can be found again and recovered for artificial incubation. The upper side of each egg is marked with a small pencilled cross. The eggs are then embedded in damp sand in a small plastic box that is then placed inside a Type I incubator.

With day temperatures from 30 to 35° C and night temperatures from 20 to 22° C, incubation lasts for 53 to 56 days. At that time the nearly developed young will slit the fragile egg shell open with a special "egg tooth." It requires another 2 to 3 days for the hatchling to finally leave the shell. The newly hatched turtles are 23.5 mm long and 20 mm wide.

They still have a yolk sac, which will be absorbed during the following 7 days. The young can be transferred into aquaria with filters and a sand bottom and also a small land section. After 4 weeks the young turtles will feed on tubifex, mosquito larvae, daphnia, cyclops, small fishes, and small pieces of raw meat.

Geochelone elegans—Indian Starred Tortoise
(Family Testudinidae)

Description: Size to 25 cm. Carapace with radiating yellow markings against a dark brown to blackish background coloration. Head and extremities yellow-brown with darker brown spots. Male with concave plastron.

Distribution and habitat: Pakistan, India, and Sri Lanka. Grassy thickets and forests with low vegetation in lower reaches of hilly and mountainous regions.

Care: Food: Vegetables, fruit, earthworms, mealworms, lean raw beef; lettuce is favored by some keepers but frowned upon by others. Should be kept in arid terrarium with an air temperature from 23 to 27° C and localized substrate heat (by means of floor heating) to 32° C. The substrate should consist of sand, and there should be a small water bowl. The terrarium should be sprayed with water in the morning.

Behavior and breeding: The starred tortoise is active primarily during the day. Occasionally it can be seen taking a sun bath, but it tends to avoid prolonged exposure to direct sunlight. Relatively little is known about the reproductive behavior of this species, although it has been bred in captivity. For copulation the male mounts the female, supporting himself with his forelegs on the carapace of the female. In contrast to the behavior in many other tortoises, the female is neither bitten nor rammed during courtship.

As early as two months prior to laying the eggs the female will start taking less food and must have more water. Preliminary attempts to dig a pit with the hind legs are also made during this period. The actual shovelling is done by alternate movements of the hind legs with the claws curved in and under. In this way the female excavates a pit that is about 10 cm deep and 6.5 cm wide; here she deposits 1 to 3 eggs (41-46 mm long and 27-32 mm wide). The eggs are covered with a viscous mucus and descend into the pit on a mucous thread. The pit is covered after the eggs are laid.

The eggs should be carefully uncovered and their upper sides marked. They are cautiously removed from the nest and placed into a plastic box filled with damp peat moss. This in turn is kept in a special brood container (Type I). At a temperature of 26 to 30° C and a relative humidity of 80%, the young turtles hatch in about 107 days. At birth they are 3.5 cm long. Food is taken for the first time on the second day. The diet should be varied and include lettuce, fruit, and mealworms. Sexual maturity is attained after about 3 years.

Geochelone pardalis (African Leopard Tortoise) (to 68 cm; central to southern Africa; arid regions with sparse vegetation) can be kept and bred under the same conditions. The incubation period lasts 90 to 92 days at 29-30° C.

Kinosternon bauri—Striped Mud Turtle
(Family Kinosternidae)

Description: Size to 12 cm. Carapace yellowish brown to blackish gray with three yellow longitudinal stripes. Plastron yellowish to olive, anterior and posterior sections movable. Two indistinct yellow lateral bands or mottling on each side of the head. Males smaller than females.

Distribution and habitat: Southern Georgia and all of peninsular Florida. Standing or slow-flowing waters.

Care: Food: Fish, squid, shrimp, mussels, water snails, worms, and pieces of meat. Should be kept in an aquarium at 17 to 34° C. Sand makes a good substrate, and roots, branches, and rocks to haul out on should be provided. Give a land section with sand for depositing the eggs. Water plants such as *Elodea, Ceratophyllum,* or similar species can be added. Over-wintering at 7 to 13° C in water or a damp substrate is recommended for adult specimens.

Behavior and breeding: This species is active primarily during the day. During courtship the male approaches a female with his neck extended and touches the female's cloacal region. The female then attempts to move away but the male follows her, nudging and sniffing her in the region of the bridge, and executing upward motions with his head angled to one side. If the female then remains motionless (this is a sign of mating readiness), the male then climbs onto her carapace, grasps the edge of her carapace with his claws, and moves his tail under that of the female. Then, with his neck extended, the male strokes the head and neck of the female with his chin barbels, while she attempts to bite into one of his limbs. During copulation, which lasts 10 to 40 minutes, the male can release all of his limbs from the carapace of the female and remain anchored only with his penis.

A few weeks later the female will bury 1 to 7 eggs about 3 to 7 cm deep in a moist sand pit. These eggs are 28.3 to 31.5 mm long and 18.5 to 19.5 mm wide. They weigh 8 grams and have a hard calcareous shell. When uncovered by the keeper they must be properly marked and transferred into a mixture of damp peat moss and sand that is placed inside a Type I brood container for incubation. After an incubation period of about 119 days (at a temperature of 25 to 30° C) the young turtles hatch. At that stage they are 24 mm long and have a weight of 3 grams; they can be raised in an aquarium on the same food as for adults. If they become very aggressive to each other they may be reared individually. Sexual maturity is reached in 5 to 7 years.

The following mud turtles can be bred under identical conditions:

Kinosternon leucostomum
Maximum size: 14 cm.
Distribution: Mexico to Ecuador.
Clutch size: 4.
Incubation period and hatchling size: 120-250 days at 24-30° C; 3 cm.

Kinosternon subrubrum—Eastern Mud Turtle
Maximum size: 13 cm.
Distribution: Southern USA.
Clutch size: 6.
Incubation period and hatchling size: 106 days at 27-30° C; 2.5 cm.
Remarks: Sexual maturity at 5-7 years.

Staurotypus salvinii—Pacific Giant Musk Turtle; Crucilla
(Family Kinosternidae)
Description: Size to 30 cm. Carapace olive brown with three longitudinal bony keels. Plastron small, connected rigidly to carapace. Orange-yellow spots on a blackish head; jaws yel-

222

lowish. Male with longer tail and heavier cloacal region.

Distribution and habitat: Pacific coast of Central America from southern Mexico to southern Guatemala and El Salvador. On the bottom of standing or slow-flowing waters in the coastal lowlands.

Care: Food: Snails, worms, fish, shrimp, freshly killed pink mice, pieces of meat, and occasionally bananas. Should be kept in an aquarium with a water level of 20 to 30 cm and a water temperature of 25 to 30° C. The land section should be removable (e.g., a plastic container with damp sand) as it is used mainly for depositing eggs. Provide submerged roots, branches, and pieces of wood in the water section.

Behavior and breeding: This species is essentially aquatic. It exhibits a characteristic defense behavior toward all potential and/or actual enemies, which has to be kept in mind when the tank is being cleaned and during feeding. First the head is pulled back into the shell, then it is suddenly thrust out with the mouth wide open, attempting to bite.

Copulation is preceded by mating foreplay that synchronizes the breeding activities of both partners. Once the male and female notice each other in the aquarium (which should be as large as possible), they remain initially motionless. If the female is ready to mate she remains motionless, pulls in her legs, and with slight upward and downward motions of one foreleg creates a slight water current in the direction of the male. Presumably specific scents (pheromones) are being released from a scent gland in front of the bridge and so directed toward the male. His reaction is usually also to release certain scents. Then he swims around the female and approaches her with snake-like head and neck motions. Sometimes the male "nods" his head or "waves" with it in side to side motions. He smells at the bridge region of the female and so presumably picks up traces of secretions from her scent gland. Apparently this is the triggering mechanism for the following behavior leading to copulation.

The male attempts to mount the female from behind, but biting from the female tends to discourage the initial attempts. However, after awhile the male supports himself with his forelegs on the carapace of the female and with his hind legs grasps the posterior edges of the shell. The male now bends his tail underneath that of the female, grasps the tail of the female between his hind legs, and inserts his black penis into the cloaca of the female. During the following copulation, which may last from 15 to 45 minutes, the female makes stationary walking movements, raising and lowering the anterior part of the body in rhythmic sequences. Then the female turns 180 degrees in the horizontal plane and makes ritualistic biting moves toward the head of the male. When retracting her head, the female opens and closes her mouth several times, presumably in order to direct a water current toward the mouth and nose of her partner. The same lower jaw clapping can also be seen in the male. The female signals the end of her mating willingness through forward movements and by trying to rid herself of the male on top. The male dismounts from the female in a lateral direction, and both animals remain facing each other for a short period, nodding with their heads and clapping their lower jaws. From then on it is particularly the female that tends to avoid meeting the male again. Therefore, it may be better and more effective if the animals are separated after mating.

When the female begins to wander around the aquarium restlessly, refuses to feed, and frequents the land section more and more, egg-laying is imminent. After examining the substrate, the female begins to dig with her forelegs at a site with moist sand. Then the pit is further wetted with water from the cloaca. Digging continues with alternate shovelling movements of the hind legs until a pit 12 to 15 cm deep has been excavated. Into this pit the female lays 3 to 10 eggs (35 x 22 mm). Finally she fills the pit up again and smoothes it over so that it is no longer noticeable. The eagerness to mate is particularly pronounced in the female just before laying her eggs. Up to three clutches of eggs can be produced per year.

The eggs, with their upper side properly marked, are transferred to a sterile layer of peat moss and incubated in a Type I brood container. The incubation period is about 145 days at 25 to 30° C. When hatching, the young turtles are about 4 cm long. They should be placed in rearing tanks—plastic containers 35 x 17 cm, with a water level at 2 cm—kept at 25 to 28° C and containing a few floating plants such as *Ceratopteris*. The substrate should be made up of a layer of sand. The young are mostly nocturnal. They should be given a diet of snails, daphnia, earthworms, small fishes, and small pieces of beef.

The following aquatic turtles can be bred under identical conditions:

Claudius angustatus (Kinosternidae)
Maximum size: 15 cm.
Distribution: Eastern Mexico to Belize.
Habitat: Aquatic.
Clutch size: 30.
Incubation period and hatchling size: 90 days at 25° C; 3 cm.
Remarks: Very large aquaria for adults. Will bite.

Chelodina longicollis (Chelidae)—**Australian Longneck**
Maximum size: 30 cm.
Distribution: Eastern Australia.
Habitat: Flowing waters with dense vegetation.
Clutch size: 24.
Incubation period and hatchling size: 63-70 days at 29 to 31° C; 2.5 cm.

Pelomedusa subrufa (Pelomedusidae)—**African Sideneck**
Maximum size: 30 cm.
Distribution: Africa south of the Sahara, Madagascar.
Habitat: Standing waters.
Clutch size: 15.
Incubation period and hatchling size: 51 days at 30° C; 2.5 cm.

Trionyx sinensis (**Trionychidae**)—**Chinese Softshell**
Maximum size: 40 cm.
Distribution: Southeastern Asia.
Habitat: Aquatic.
Clutch size: 40.
Incubation period and hatchling size: 54–73 days at 22-32° C; 2.5 cm.
Remarks: Very large aquaria with sand as bottom substrate for burrowing (for adult specimens).

Chelydra serpentina (Chelydridae)—**Common Snapping Turtle.**
Maximum size: 47 cm.
Distribution: Eastern North and Central America to northern South America.
Habitat: Aquatic.

Clutch size: 5.
Incubation period and hatchling size: 102-110 days at 26 to 29°
C; 1.7 cm.

Sternotherus minor—Loggerhead Musk Turtle
(Family Kinosternidae)
Description: Size to 11 cm. Carapace and limbs tan to black;
shields of juveniles sometimes with pattern of radially arranged lines. Plastron lighter. Male with very large head.
Distribution and habitat: Southeastern United States. On the
bottom in standing or slow-flowing waters.
Care: Food: Snails, mussels, tubifex, earthworms, aquatic insects, small fishes, shrimps, and small pieces of beef heart.
Must be kept in an aquarium with 8 to 10 cm of water at 20
to 34° C; provide a substrate of sand and an egg laying container filled with moist sand. Plant with *Philodendron* or similar types. Over-wintering is possible in a damp substrate at
10° C.
Behavior and breeding: This species is mainly nocturnal.
When endangered it will emit a foul-smelling musk secretion
that is usually sufficient to ward off any enemies. Mating is
most easily accomplished with animals that have been kept
apart for awhile. The courtship procedure is fairly similar to
that of *Staurotypus salvinii.*

Following optical and olfactory contact, the male approaches the female, smells her, touches her, and bites into
the lateral rim of her carapace. He then mounts the female
from behind—making lateral movements with his head—and
induces her to withdraw her head into her shell. Then the
male moves back slightly and establishes contact with the
female's cloaca with his tail. This puts him into a virtually
vertical position during copulation. Apart from contact via
the penis, the male holds on loosely with his hind legs to the
rim of the female's carapace.

Eggs may be laid as often as four times a year. The female
selects a section of the land area that contains adequately
moist sand. There, in a shallow pit, she buries 2 to 5 eggs
(dimensions 29.8 x 17.1 mm) and then covers the nest over
again. Once uncovered by the keeper, the eggs are marked
properly on their upper surface and then transferred into
damp peat moss. They are then placed inside a brood container (Type I). The first young (about 25 mm in length)

226

hatch after 61 to 119 days at an incubation temperature of 25 to 30° C. The young turtles should be raised individually in plastic containers with an area of not less than 20 square cm decorated with roots or similar materials to provide hiding and climbing opportunities and with sand as a bottom substrate. They feed on the same foods as their parents.

Sternotherus odoratus (Stinkpot) (13 cm; eastern USA; aquatic) can be bred under identical conditions. The clutch consists of up to 6 eggs and the incubation period lasts for 103 to 132 days at 20 to 30° C. Upon hatching the young are 2.3 to 2.5 cm long. Sexual maturity is reached after 2 to 4 or more years.

Terrapene carolina—Common Box Turtle
(Family Emydidae)
Description: Size to 20 cm. Carapace highly arched, dark brown to black, often with yellow spots, patches, or radiating lines. Plastron light to dark brown without radiating lines. Hind feet without distinct webs. Male frequently with red iris, females frequently with yellow-brown iris.
Distribution and habitat: Southern Canada to northern Mexico. Forests and meadows, usually in the proximity of standing water.
Care: Food: Snails, insects, worms, fish, fruit, vegetables, beef heart, and canned cat foods. Should be kept in a semi-moist terrarium or outdoor terrarium with sunlight at 20 to 28° C. The substrate can consist of a damp peat moss and sand mixture, in depth equal to height of carapace of largest animal. The shallow water section should have roots and branches as hiding places (plants and other decorative arrangements on the ground are usually destroyed by these animals). Provide a source of light. Should be sprayed with water daily. Can be over-wintered in a damp substrate at about 10° C.
Behavior and breeding: Although box turtles belong in the family Emydidae, they are primarily terrestrial, being particularly active during the morning and early evening hours. Similar to the other species of this genus, the box turtle also has a characteristic defensive behavior. It pulls in the head and extremities and then the middle and anterior sections of the hinged two-part plastron are pressed so firmly against the edges of the carapace that it is impossible to remove the ani-

mal from this "box." This mechanism provides effective protection against all enemies, except automobiles (box turtles suffer a high death toll near highways).

The mating season for this species can extend from spring to fall. Courtship is initiated by the male with a series of optical, tactile, and olfactory signals. Usually the male approaches the female frontally, stops a few centimeters in front of her head, elevates his carapace on all four legs, and extends his head and neck in the direction of the female. She in turn pulls her head into the shell, remains stationary, and observes the male. The male then walks around the female while making jerking forward and backward motions with his shell and his head, at the same time bumping against the female's carapace. He also sniffs at the female and bites at random at her soft skin parts. This behavior can last for hours until the male succeeds in climbing onto the carapace of the female from behind. The front legs are then supported on the female's back and his hind legs are brought underneath the posterior end of the female's carapace.

At the same time, while lying on the back of the female's carapace, the male continually tries to bite the female's head and his own head and neck are extended far forward. This is to persuade the female to open the closed posterior section of her plastron. Once this has been accomplished, the male pushes his feet between carapace and plastron of the female and wraps his tail around hers so that both cloacal openings are facing each other. During the following copulation the female closes the posterior end of her plastron against the carapace, so that the hind legs of the male become trapped. With his front legs the male then kicks himself off the carapace of the female until his carapace actually touches the ground, and then returns into a vertical position. This movement is repeated with every penis thrust.

A few weeks later the female starts digging a nest hole. At first she uses her front legs, then also her hind legs, until a depth of about 11 cm is reached. She lays 3 to 8 white elliptical eggs that are 32 x 19.5 mm. She then closes the pit again with alternate shovelling movements of the hind legs and smooths the pit over again with her plastron.

The eggs have to be dug up very carefully, and the tops should be marked clearly. They are then transferred into a container filled with loose peat moss that is placed inside a

Type I brood container. At an incubation temperature of about 28° C, the first young turtles will—with the aid of their special egg tooth—burst the egg shell after about 70 days. However, they will remain inside the shell for a few days longer before they completely hatch. The young turtles are about 20 mm long and have a blackish gray plastron and similarly colored legs. Initially they still have a large yolk sac, which is resorbed during the following two weeks. In order to avoid late hatchlings being damaged by earlier ones, it is advisable to separate all hatchlings into individual clear plastic containers with damp peat moss and fresh moss. Soon the young turtles will begin to dig into this substrate. The first food is taken at the age of 3 weeks. It should consist of mosquito larvae, raw fish, earthworms, tubifex, and similar items. Sexual maturity is reached in 5 to 7 years.

The following terrestrial emydid species can be kept and bred under identical conditions:

Rhinoclemys pulcherrima
Maximum size: 20 cm.
Distribution: Central America.
Habitat: Rain forest.
Clutch size: 3.
Incubation period: 100-147 days at 27° C; 5 cm.

Geoemyda spengleri
Maximum size: 13 cm.
Distribution: Southeastern Asia.
Habitat: Damp mountain forests.
Clutch size: 3.
Incubation period and hatchling size: 74-110 days at 27° C; 3 cm.

Testudo graeca—Mediterranean Spur-thighed Tortoise
(Family Testudinidae)
Description: Size to 25 cm. Carapace yellowish to olive with dark patches; plastron yellow-brown with dark patches. Distinct spur on upper thigh; usually only one tail shield; tip of tail without large scale. Male with longer tail and more concave plastron.
Distribution and habitat: Eastern Balkan Peninsula; south of

the Danube to Macedonia and European Turkey; islands of Thasus, Samotraki, Euboea, and Sicily; southern Spain to North Africa; Asia Minor and Middle East to Iran. Semi-moist to arid regions with bush vegetation.

Care: Food: Fruits, vegetables, wild flowers such as dandelion, chickweed, clover, lupine, lamb's lettuce, knot-grass, and others, lean chopped beef, earthworms, hard-boiled egg, and rice.

Should be kept in a dry terrarium or in an outdoor terrarium, the temperature during the day 20 to 32° C, slightly cooler at night. The substrate can consist of sand, flat smooth rocks (with floor heating to provide warmth), and larger rocks to form a cave; provide a small water bowl. Plants (which must be out of reach of the tortoises or can be omitted altogether) such as agaves, succulents, and small bushes (to provide shade) can be used. This species should be overwintered at 5 to 10° C in a mixture of damp leaves and moss after the animals have been fasted for awhile and bathed in luke-warm water several times.

Behavior and breeding: This species tends to emerge in the morning from its nightly resting place to soak up the warm morning sun. The hot noon period is spent in the shade of bushes, and in late afternoon the animal goes hunting for food once again. During habitat studies in Greece and Turkey we were able to measure the anal temperature of these animals as well as the environmental temperatures with electronic thermometers and also to measure the light intensity with a luxmeter throughout the day. Just as *Chamaeleo chamaeleon* can regulate its body temperature within a certain range, so can *Testudo graeca* in order to avoid death due to overheating. In essence, its body temperature on a cool morning is higher than the air temperature and increases with increasing air temperature, yet it always remains below the maximum values of a hot summer's day. During the evening the body temperature falls together with a lower environmental temperature but remains slightly above ambient temperature during the night.

Males test their strength in ritualistic fights during the courtship season. This consists of biting each other on the extremities, elevating the body high on all four legs, and—with the head pulled into the shell—ramming the carapace of the opponent. Sometimes males attempt to maneuver their carapace under that of the opponent in order to turn him over on to his back. However, if a female appears one of the males will move quickly toward her, making a frontal approach while displaying jerking lateral movements of the extended head and neck. He then smells the female and bites her on the head or the extremities. If the female keeps moving the male will follow, attempting to stop her and get her to pull her head into the shell by ramming her laterally or by biting her. Once this has been accomplished the male will climb onto the back of the female's carapace, supporting himself with his front legs and wrapping his tail around that of the female. This is followed by copulation, during which the male gives off a continuous high-pitched chirp.

A few weeks after copulation the female will dig a pit and lay 2 to 8 eggs 32 to 42 mm long with a diameter of 22 to 32 mm. The eggs should be dug up, their upper side properly marked, placed in damp peat moss, and finally transferred to a Type I brood container. The young turtles (about 5 cm at hatching) will hatch in about 78 to 84 days at a temperature of 29° C and a relative humidity of 70%. Due to the strongly bent position inside the egg, the hatchlings have a longitudinal fold in the soft plastron that tends to smooth out and grow over during the following days. With a varied diet and regular UV or solar radiation (important during the juvenile stage), this species reaches sexual maturity in about 5 years.

The following species can be kept and bred under identical conditions:

Testudo hermanni (**Testudinidae**)
Maximum size: 20 cm.
Distribution: Southern Europe.
Habitat: Semi-moist to dry bush lands.
Clutch size: 9.
Incubation period and remarks: 62-66 days at 27-30° C; sexually mature in 7 years.

Testudo marginata (**Testudinidae**)
Maximum size: 30 cm.
Distribution: Southeastern Europe.
Habitat: Dry regions with sparse vegetation.
Clutch size: 14.
Incubation period: 59-73 days at 29 to 30° C.

Agrionemys horsfieldi (Testudinidae)—Iranian Tortoise

Maximum size: 28 cm.

Distribution: Pakistan, Afghanistan, Iran, and adjacent USSR.

Habitat: Dry regions.

Clutch size: 5.

Incubation period and remarks: 55-77 days at 30 to 34° C; sexually mature in 10 years.

Terrapene ornata (Emydidae)—Western Box Turtle

Maximum size: 14 cm.

Distribution: Great Plains and southwestern USA.

Habitat: Sandy and semi-dry regions.

Clutch size: 8.

Incubation period: 60-70 days at 28 to 30° C.

REPTILES
Lizards (Sauria)

Acanthodactylus boskianus—**Fringe Finger**
(Family Lacertidae)
Description: Size to 20 cm. Dorsum yellow to orange-brown, frequently with 7 white longitudinal bands and light to dark brown spots. Ventral side white to yellow. Toes with wide scale combs. Males with enlarged cloacal region and strongly developed femoral pores.
Distribution and habitat: North Africa, Israel, southeastern Turkey, and Iran. Sandy semi-desert and desert regions with sparse vegetation; also found close to human habitation.
Care: Food: Insects, spiders, small pieces of beef heart. To be kept in a dry terrarium at air temperatures from 25 to 36° C during the day and 15 to 22° C during the night. Provide local heating with a radiator or substrate heating and also regular UV radiation. The substrate should be an at least 8-cm layer of sand; a moist section with plants is required for egg-laying. Decorations can include rocky structures and succulents. Provide a small water bowl. This lizard can be over-wintered at about 10° C.
Behavior and breeding: This species is active during the day and prefers sunlight. In order for it to survive in a hostile environment it has developed a peculiar feeding strategy. In contrast to the other members of this family, this species feeds not only on live organisms but also on dead organic debris. The question thus arises, how can *Acanthodactylus* find prey that does not provide movement stimuli? Apart from the visual sense, also important for this species are olfactory and taste perception. When moving, the lizard's tongue frequently passes over the substrate. If a piece of meat is recognized as such, it will immediately be seized with the jaws and eaten without delay. The ability to recognize dead prey allows exploration of a new food source and could be the reason why this species is more frequently found near rubbish disposal areas around human habitation than in uninhabited desert areas (e.g., the Sinai Peninsula).

Our males of *Acanthodactylus boskianus* inhabit a dry terrarium of 120 x 60 x 40 cm together with other desert species and form a hierarchy. This is especially notable while feeding, "sunbathing," and during fighting for a female. The strongest male is the "alpha male" of the group and thus is always the first one to feed on such delicacies as newly molted grasshoppers or crickets. Only after it has had enough to eat is the "beta male" permitted access to the food. However, females are excluded from the male hierarchy; a female can eat at the same time the alpha male is feeding or use the same site for "sunbathing" (under a radiator) together with a male without getting bitten or chased away.

Within a captive breeding group only the alpha male has the right to reproduce. There can be as many as 14 mating periods within one year, and if a subordinate male tries to approach a female during one of these periods it will immediately be driven off by the dominant male. The typical aggressive posture of the alpha male involves raising the body high on all four legs, inflating the throat region, and moving the head up and down while approaching the trespasser. Usually this is sufficient to intimidate the subordinate male, and a fight is avoided. The subordinate male signals his submissiveness through tail trembling and repeated alternate up-and-down motions of the front legs, and then it flees. However, if these submissive signals are not forthcoming, it will be licked by the dominant male and then bitten in the tail or the neck region.

Courtship commences with the male approaching the female with his neck bent downward and displaying an inflated throat region. He licks her body several times, moves around her in a semi-circle, and smells the base of her tail, the cloacal region, and the other side of her body. Should the female not yet be ready to mate she will threaten the male by raising her body, turning her head toward the male, and opening her mouth wide while shaking the anterior part of her body from side to side. Sometimes such an unwilling female will also push the male with her snout; however, there is no real biting involved.

On the other hand, a female willing to mate will remain motionless, with her body firmly pressed to the ground, elevating slightly the base of her tail. Apparently the male stimulates the female into copulation by pushing his head into

234

her side and finally by biting firmly into the female's tail base. While doing that he places his forelegs on the posterior end of the female, reaches around the female's tail base with one hind leg, and maneuvers his tail under that of the female. After brief attempts to stabilize their position, copulation (which may last from 1 to 4 minutes) takes place. While the female remains completely passive throughout—she even keeps her eyes closed—the male tries to fend off any approaching rivals.

About 14 days later the female starts to wander restlessly through the terrarium, frequently checking the ground with her tongue. Once the female has found a suitable site with moist fine-grained sand, she pushes her snout into the sand and then starts to shovel with alternate movements of the fore and hind legs until an approximately 8-cm-deep hole has been dug. During the following hours the female will lay up to 7 eggs in this pit and then cover it over again and compact the sand with her snout and body. Clutch size and number of clutches per year seem to depend upon the size, age, and condition of the female. Mating can be repeated at 5-day intervals.

Since the incubation conditions for these eggs (length 1.6 cm, diameter 0.8 cm) with their parchment-like shells are rarely ever optimal in a terrarium, the eggs should be unearthed, their upper sides marked properly, and then transferred into plastic containers with damp fine-grained sand. The containers are then placed inside a regular (Type I) brood container. The yellow and black banded juveniles (8.5 cm long) will hatch after 89 to 100 days at a temperature of about 28° C. Once the umbilical cord has dried up the young are placed into a dry terrarium with fine sand as a substrate (cover half of the area with damp sand and the other half with dry sand). Soon they will start excavating small sleeping hollows just like those of their parents. It is imperative that they be given a daily water spraying and regular UV radiation (we have UV-A tubes in continuous operation) together with a calcium- and vitamin rich diet.

Sexual maturity is attained after 1½ years. As with all juvenile reptiles from the temperate and subtropical zone, overwintering can be omitted for this species during the first year of their life.

The following species can be kept and bred under identical conditions:

Acanthodactylus erythrurus
Maximum size: 23 cm.
Distribution: Spain, Morocco, Algeria.
Habitat: Arid regions with sandy to rocky substrate.
Clutch size: 7.
Incubation period: 35 to 40 days at 30° C; 6-8 cm.
Remarks: Sexual maturity in 2 to 3 years.

Eremias strauchi
Maximum size: 20 cm.
Distribution: Armenia, Transcaucasus, central Asia, north-eastern Turkey, Iran.
Habitat: Steppe, semi-desert.
Clutch size: 3.
Incubation period: 49 days at 20 to 30° C.

Agama impalearis—**Atlas Agama**
(Family Agamidae)
Description: Size to 30 cm. Dorsum gray-brown; in excited or alarmed males the color changes to light blue with a yellow dorsal band; females yellow with blue-red spots. Ventral side light. Males with strongly developed pre-anal pores.
Distribution and habitat: Morocco to western Sahara, Tunisia, Algeria, Niger (Air Mountains). Rocky deserts, on rocks.
Care: Food: Insects, newly-born mice or hamsters, sweet fruits, dandelion leaves. Should be kept in a dry terrarium at an air temperature from 28 to 40° C during the day and 22° C at night; also supply localized (spot) heating with radiator or floor heating, plus regular UV radiation. Mist in the morning. Use sand as substrate. If decorations such as rocky structures and branches are given, then planting with succulents or similar species can be omitted. Supply a small water dish.
Behavior and breeding: Males are strongly territorial and defend their territory against conspecifics. If another male approaches,the defender signals his territorial claims by displaying his throat pouch, inflating the body, elevating himself on his forelegs, and making jerky up and down motions of the head and upper body. Simultaneously the defending male's back turns light blue and a yellow center line appears. If at that stage the intruder does not flee immedi-

ately, a fight ensues, with both males stalking around each other, threatening and tail-whipping.

During the mating season the male approaches a female while nodding his head. If the female arches her back, elevates her body on all four legs, and raises her tail slightly, the male will encircle her while continuing his head nodding. Finally the male climbs on the back of the female, grabbing her neck with his jaws and reaching with one hind leg for the base of the female's tail. Then he maneuvers his tail under that of the female and presses his cloaca against hers. Copulation commences with the introduction of the hemipenis.

A few weeks after mating the female deposits 5 to 20 eggs in a pit she dug herself. The eggs are white, 12 x 22 mm in size, and weigh 0.6 grams. They should be uncovered, their upper sides properly marked, and then transferred into a plastic bowl with damp sand. This bowl is then placed inside a Type I brood container. The young hatch after 46 to 54 days at an incubation temperature of 30° C. They are 78 to 81 mm long and weigh approximately 1.2 grams. They should be transferred to a dry terrarium and be given a diet of small crickets, grasshoppers, and similar insects. They should occasionally be given calcium and vitamin supplements as well as regular UV radiation.

The following species can be kept and bred (in a large terrarium) under identical conditions:

Agama atricollis (**Agamidae**)
Maximum size: 30 cm.
Distribution: Southern and eastern Africa.
Habitat: Savannah.
Clutch size: 10.
Incubation period: 74-75 days at 25 to 30° C.

Agama caucasica (**Agamidae**)
Maximum size: 35 cm.
Distribution: Throughout the Caucasus, southern and north eastern Turkey, Iraq, Afghanistan, Iran, Pakistan.
Habitat: Rocky dry regions with sparse vegetation.
Clutch size: 14.
Incubation period: 53-57 days at 28 to 29° C; 7.2 cm.
Remarks: 5-8° C for over-wintering.

Agama sanguinolenta (Agamidae)
Maximum size: 30 cm.
Distribution: Eastern Caucasus to central Asia, northeastern Iran, northern Afghanistan, to northwestern India.
Habitat: Dry areas with sparse vegetation.
Clutch size: 11.
Incubation period: 50-52 days at 27 to 31° C; 3.5 cm.

Agama stellio (Agamidae)
Maximum size: 35 cm.
Distribution: Corfu, northern Greece, Middle East, Israel, Sinai Peninsula, Iraq, Cyprus, lower Egypt, Arabian Peninsula.
Habitat: Stone walls, trees, dry rocky region with sparse vegetation.
Clutch size: 10.
Incubation period: 52-55 days at 28-29° C.
Remarks: Sexual maturity in 1½ years.

Amphibolurus nuchalis (Agamidae)
Maximum size: 25 cm.
Distribution: Australia.
Habitat: Dry regions.
Clutch size: 6.
Incubation period: 75-79 days at 27° C; 7 cm.

Uromastyx acanthinurus (Agamidae)
Maximum size: 40 cm.
Distribution: Northern Africa.
Habitat: Rocky desert.
Clutch size: 18.
Incubation period: 11 days at 28 to 35° C; 7.3 cm.
Remarks: Higher incubation temperature (35° C) more favorable. Sexual maturiy in 3-4 years. Diet should include fruit, dandelion flowers, and similar plants.

Leiocephalus carinatus (Iguanidae)
Maximum size: 25 cm.
Distribution: Cuba.
Habitat: Rocky slopes.
Incubation period: 51-79 days at 27° C.
Remarks: For diet see *Uromastyx*.

Sauromalus hispidus (Iguanidae)—Mexican Chuckwalla
Maximum size: 40 cm.
Distribution: Western Mexico.
Habitat: Dry, rocky areas.
Clutch size: 21.
Incubation period: 94-96 days at 29-31° C; 13.6 cm.
Remarks: Diet as in *Uromastyx*.

Sauromalus obesus (Iguanidae)—Chuckwalla
Maximum size: 43 cm.
Distribution: Southwestern USA into Mexico.
Habitat: Dry bushland, rocky regions.
Clutch size: 13.
Incubation period: 79 days at 35° C; 5.7 cm.
Remarks: Diet as in *Uromastyx*.

Cnemidophorus sexlineatus (Teiidae)—Six-lined Racerunner
Maximum size: 30 cm.
Distribution: Southeastern and central USA.
Habitat: Dry bushland, paddocks, forests.
Clutch size: 6.
Incubation period: 61-75 days at 25-32° C.

Amphibolurus barbatus complex—Bearded Dragon
(Family Agamidae)
Description: Size to 53 cm. Dorsum grayish brown with lighter patches; ventrally white with black spots. Throat and side of head with spine-like extended scales.
Distribution: Australia. Semi-deserts and savannahs with plant growth.
Care: Food: Insects, spiders, newly born mice, fruit, vegetables, lettuce, dandelion leaves and flowers. Should be kept in a dry terrarium at an air temperature from 28 to 39° C during the day and at 18 to 24° C during the night. Localized heating for basking as well as regular UV radiation should be provided. Substrate should consist of sand (area with damp sand at least 20 cm deep is required for egg-laying). Some climbing facilities, such as acacias and some rock formations, required. Plants can include *Rubus, Sansevaria,* or similar species, but plants can also be omitted. A water dish for bathing is required.

Behavior and breeding: A hierarchy among bearded dragons tends to develop very early, possibly within days after hatching. Initially this hierarchy is dependent upon size and weight and becomes very conspicuous during feeding. When food such as small crickets, grasshoppers, or black beetles is offered the larger bearded dragons feed first while the smaller ones merely watch with their tails bent upward in excited anticipation. The presence of the larger siblings obviously acts as an inhibitor to the feeding urge. Only after the larger animals have had their fill do the smaller ones get to feed. During the first 10 months young bearded dragons develop a deferential social behavior pattern. The strongest and most aggressive animal within a group occupies the alpha position. If this animal selects a particular site for sunbathing, the previous occupant will move away. From that position the leading male, his tail slightly curved upward, then surveys his territory. If the alpha male notices a fight between two subordinate animals, it will run over to them, licking one of the two and nodding its head excitedly 3 to 5 times. In view of this dominant gesture the two immediately cease fighting. They execute circular motions alternately with each leg, the so-called "arm turning." This ritualistic response pacifies the dominant animal and so prevents further aggressive action.

If another male is introduced into an established group, it will initially be observed closely by all other group members, their tail tips slightly bent upward. If the new male nods its head several times, the alpha male will run toward the newcomer. The beard and tip of tail turn black within seconds, and the jet black beard is spread in a threatening manner. The alpha male nods his head and attempts to intimidate the opponent by enlarging his body profile, positioning the body low and obliquely toward the opponent. If the newcomer neither flees nor gives an "arm turning" appeasement signal but instead nods his head and moves the tip of his tail excitedly, a fight will break out. One animal will run halfway around the other, biting into the tail of the opponent. Alternatively, the animals try to intimidate each other with threatening postures, encircling the opponent in an oblique position and attempting to jump on top of each other. If one of the opponents succeeds he will bite firmly into the thorny extended neck scales so that the suppressed animal can barely move its head to defend itself. If the overpowered animal presses its

240

DOMINANCE

THREAT

FIGHTING

APPEASEMENT

DEFENSE

SUBMISSION

COURTSHIP

RECOGNITION

Social behavior in *Amphibolurus barbatus*.

head and body flat against the ground and so signals his submissiveness, the winner will turn away. In spite of these aggressive interactions, which occur during hierarchy conflicts and territorial defense and which involve severe biting, there are no serious injuries, so this must be considered more as ritualistic fighting.

Bearded dragons become sexually mature in one to 2 years. During the following reproductive period lasting about 4 months only the alpha male will mate with a female of a particular group. The male approaches the female, nodding his head. Should the female be willing to mate she will press her body to the ground and raise the tail slightly. The male then climbs on the back of the female and bites solidly into one side of the neck region. In doing that he reaches with one hind leg for the tail of the female and presses his cloacal region against hers.

Shortly before the eggs are laid the female ceases feeding. About 5 days before the eggs are actually laid, the female begins to pace through the terrarium, scratching at various places in search of a damp area in the substrate. Then she begins to dig, at first with the forelegs and then with the hind legs, excavating a cave just large enough for the female to fit into. During the next 45 minutes the eggs are deposited and subsequently the entire clutch is covered over again with sand. This still loose sand is compacted with the feet and head so that there is nothing to indicate that eggs have been laid there. A female can lay from 15 to 27 eggs in five intervals about 2 to 4 weeks apart during one reproductive cycle. Since this is a substantial energy consumption, the female must subsequently be given special supplemental foods as well as calcium and vitamins.

Egg-laying terminates the involvement of the parents in producing their progeny. The eggs are white and have a parchment-like shell (on the average they are 3 cm long with a diameter of about 1.8 cm and a weight of 7.1 to 8.4 grams). They should be carefully uncovered and their upper sides clearly marked with a pencil in order to ensure correct re-positioning of the eggs after transfer; incorrect positioning could kill the embryo. The eggs are then placed into a Type I brood container.

The incubation period ranges to about 91 days at 25 to 27° C. At hatching the young bearded dragons have a size of 7.5

to 8.5 cm. They possess a barely visible (0.5 mm long) egg tooth with which they slice the tough egg shell open. They force their body out of the shell through powerful rotational movements during the following few hours. At that stage the young should be transferred to a dry terrarium. From the third day on they will feed on small crickets, black beetles, and similar insects. In view of the rapidly developing hierarchy, the last animals to hatch are best fed separately from the already larger first-hatched ones. It is particularly important that during the prime growth period the young lizards receive sufficient calcium in the form of crushed cuttlebone and regular UV radiation as prophylactics against rickets. We found the UV-A fluorescent tubes in continuous use or 5-minute exposures at 1 meter to Osram-Ultra-Lux-type bulbs twice a week to be quite satisfactory.

Anguis fragilis—Slowworm
(Family Anguidae)
Description: Size to 50 cm. Dorsum blue-gray to bronze-brown, frequently with a black dorsal band. Male frequently with blue spots, the ventral area black or gray. Snake-like in appearance; however, the eyes have movable lids. Smooth scales.
Distribution: Throughout Europe, Asia Minor, the Caucasus, to Iran. Forests, slopes with low brushland, semi-shaded meadows, parks.
Care: Food: Worms, slugs, sometimes also grasshoppers or crickets, small "moving" pieces of beef heart, and spiders. Should be kept in a semi-moist terrarium at 20 to 28° C during the day and 15 to 18° C at night. Spray daily with water. Localized heating with radiators. A substrate of leafy forest soil or a mixture of sand and peat moss covered by a layer of fresh moss is good. Pieces of bark or roots make usable decorations. Plants can include ferns, wool grass, moneywort, or similar species. Provide a small water dish. Overwintering at about 5° C in a damp peat moss/leaf mixture is recommended.
Behavior and breeding: The ground-dwelling slowworm spends the daylight hours mostly hidden under decaying toppled trees, underneath wood piles, or in rodent burrows. Sometimes they even dig their own burrows, head-first, into

loose soil. Although this species has the occasional "sun bath," it really only becomes active at dusk.

Slowworms are color-blind and they find it difficult to distinguish shades of gray. Therefore, they use primarily their sense of taste and smell when hunting for prey. While searching for food the animal stops very frequently, raises its anterior end slightly, flicks the dark blue tongue, and vibrates the tips of the tongue. Once prey has been sensed, the tongue begins to flick at a rapid pace and the animal elevates its head and thrusts forward the head with its jaws wide open. The prey is grabbed, bashed against the ground several times to knock it out, and then swallowed.

Slowworms display a special defensive mechanism when handled. They attempt to frighten off the attacker by violently thrashing their body from side to side and giving off foul-smelling excrement, thus trying to gain their release.

Slowworms are solitary for most of the year. Only during the winter months do they tend to congregate in groups of up to 30 in order to hibernate together in protected ground burrows or under tree roots. Such winter accommodations are often also sealed off with a soil "plug." Come spring the animals re-emerge, and soon thereafter mating begins.

During the courtship the male follows the chosen female for hours until he succeeds in biting firmly into one side of the head or neck. He then wraps his body around hers until there is mutual cloacal contact. Insertion of the hemipenis is followed by sperm transfer.

About 3 months after copulation the female gives birth to 5 to 26 young (our own average was 11 young). As with all ovoviviparous reptiles, at birth the young are still inside a gelatinous transparent yellow egg membrane. They burst out of this membrane with a sudden powerful body movement. These young are 7 to 9 cm long, light gray in color, and have a dark median dorsal band. In order to monitor growth and condition of the young more adequately, they should be transferred to a separate semi-moist terrarium. Soon thereafter the first food is taken in the form of small worms and snails. They molt about 4 times a year. The old skin slides together into rings and then slips off posteriorly.

Once the young *Anguis* have grown to a size of about 25 cm (in the wild this would be at an age of 4 years), they become sexually mature.

Anolis carolinensis—**Green Anole, American Chameleon**
(Family Iguanidae)
Description: Size to 22 cm. Dorsum light green to brown with lighter serrated median dorsal line. Ventrally white. Male with large pink throat fan. Male with enlarged postanal scales.
Distribution and habitat: Southern USA from Texas to Virginia. Trees, bushes, fences.
Care: Food: Insects, spiders. Should be kept in a semi-moist terrarium at air temperatures from 23 to 30° C during the day and 18 to 20° C during the night. The terrarium should be sprayed with water morning and evening. Localized heating with a radiator plus regular UV radiation is recommended. The substrate should consist of leafy soil or a mixture of peat moss and sand covered with green moss. Climbing facilities must be provided in the form of branches or an epiphyte-covered log, *Bilbergia, Aechmea, Tillandsia, Cryptanthus, Scindapsus,* or similar plants species.
Behavior and breeding: The males establish firm territories, not only in the wild but also in a large, densely planted terrarium. One frequently can find them in elevated positions claiming their respective territory with a threatening display of head nodding and inflating their throat fan. This involves raising and lowering the head several times in succession and moving the anterior part of the body in the same rhythm. This display is further enhanced by the semi-circular extrusion of the pink throat flap or dewlap supported by the hyoid bone. If another male approaches, the resident male first tries to intimidate the other male with a threat display; it stares at the intruder, turns chocolate brown, and then goes through its ritualistic display behavior. It raises the dorsal comb, flattens its body, positions itself laterally, and stalks toward the intruder, the body elevated on all four legs. If the intruder does not leave and instead also resumes a threatening posture, a fight will ensue. The opponents encircle each other and try to bite into each other's snout or dorsal comb.

In the wild defeated males have sufficient opportunity to avoid further fights with dominant males. However, in a terrarium special care must be taken to provide sufficient sunbathing and feeding sites and sufficient common space for subordinate animals outside established territories. Small terraria should only be used for individual pairs of anoles.

The breeding season of this species extends over a period of 4 to 5 months. During the courtship a male, his throat fan spread and with his head and anterior body repeatedly "nodding," approaches a female. The male follows the fleeing female, runs parallel to her, keeps on nodding but at a more rapid frequency, and tries to bite laterally into her neck or shoulder region. If the female is willing to mate she stops and bends her neck region downward. The males bites into it, maneuvers his body under that of the female, and following cloacal contact copulation takes place. This may last about 6 minutes.

From that point on the female becomes increasingly heavier and less active. About 14 days after mating the female starts looking for a warm and damp site on the ground. There she begins to dig a hole with her head and forelegs, into which she deposits 1 or 2 eggs. If the eggs miss the hole the female will push them in with her head. After the eggs have been laid the hole is filled in again with substrate.

The eggs can either be left in the terrarium (the particular site will have to be kept moist, though) or be removed for better control (the upper side must be properly marked) and placed into a Type I brood container. The most suitable substrate is a damp mixture of peat moss and sand or swamp moss (sphagnum). The young will hatch in 35 to 40 days at 30° C. At hatching they are 50 to 60 mm long. They should be transferred into a small semi-moist terrarium such as a plastic container with wire gauze cover and lateral ventilation. The rearing temperature should be a constant 25° C air temperature. The initial diet should consist of fruitflies, leaf lice, and small crickets. In addition, vitamin-enriched honey can also be given. Regular UV radiation is recommended. After 7 to 8 weeks the young anoles will have doubled their hatching length. At that stage the males begin to "display" and establish their territories. Now the time has come to separate them; they should be kept either in pairs in smaller terraria or in a group in a large terrarium. Sexual maturity is reached in about 9 months.

The following *Anolis* species can be kept and bred under identical conditions, the size of the terrarium and the food requirements depending upon the species concerned:

Anolis equestris—Knight Anole
Maximum size: 50 cm.
Distribution: Cuba.
Habitat: Forested regions.
Clutch size: 1.
Incubation period and hatchling size: 50-70 days at 25 to 30° C; 11.5-15 cm.

Anolis garmani—Jamaican Anole
Maximum size: 37 cm.
Distribution: Jamaica.
Habitat: Forested regions.
Clutch size: 2.
Incubation period and hatchling size: 60-64 days at 20 to 26° C; 7.5 cm.
Remarks: Sexual maturity in 9 months.

Anolis marmoratus
Maximum size: 26 cm.
Distribution: Guadeloupe.
Habitat: Rain forest, cocoa plantations.
Clutch size: 2.
Incubation period and hatchling size: 50 days at 28° C; 7.5 cm.

Anolis oculatus
Distribution: Dominca.
Habitat: Damp forests to dry regions.
Clutch size: 2.
Incubation period and hatchling size: 37 days at 28° C; 4.5 cm.

Anolis porcatus
Maximum size: 21 cm.
Distribution: Cuba.
Habitat: Forested and agricultural areas.
Clutch size: 2.
Incubation period: 60 days at 20 to 26° C.

Anolis roquet
Maximum size: 19 cm.
Distribution: Martinique.

Habitat: Tropical rain forests.
Clutch size: 2.
Incubation period: 45-55 days at 28° C.

Anolis bimaculatus sabanus
Maximum size: 17 cm.
Distribution: Lesser Antilles.
Habitat: Rocks, tree roots.
Clutch size: 2.
Incubation period: 43 days at 29-30° C.

Anolis trinitatis
Maximum size: 18 cm.
Distribution: St. Vincent.
Habitat: Brick walls, buildings, plantations.
Clutch size: 2.
Incubation period: 45-55 days at 28° C.

Anolis lineatopus
(Family Iguanidae)
Description: Size to 19 cm. Dorsal side brown with dark cross bands. Ventral side lighter. Males with yellow throat fan and neck comb.
Distribution and habitat: Jamaica. Backyard trees and parks.
Care: Food: Insects. Should be kept in a semi-moist terrarium at air temperatures from 20 to 35° C during the day and 18 to 20° C at night. For further details refer to *Anolis carolinensis.*
Behavior and breeding: This species exhibits similar territorial and reproductive behavior as the green anole. However, there can also be intraspecific agonistic interaction between females as well as between males. In the wild the young of *A. lineatopus* occupy a different ecological niche from the adult specimens. Young animals are commonly found close to the ground among weeds, unlike adults that occur primarily in trees. In this way juveniles avoid being preyed upon by the adults.

The search for food is mainly visually oriented, as has been shown experimentally. In essence then, the sense of smell is not used in locating food. Selection of prey depends largely upon individual preferences for particular food items. This can be variable and is subject to change with time.

In captivity reproduction can occur throughout the year.

The eggs are deposited at a damp site on the substrate. The clutch consists invariably of only one egg. For better control and closer monitoring, the egg should be removed and (after marking the upper side) embedded in damp moss or in a damp mixture of sand and peat moss (brood container Type I).

The incubation period lasts from 53 to 60 days at a temperature of 25° C. Rearing is the same as for *Anolis carolinensis*. This species becomes sexually mature in 8 to 16 months.

Basiliscus basiliscus—American Basilisk, Brown Basilisk

(Family Iguanidae)

Description: Size to 90 cm. Dorsum greenish to olive with brown cross bands. A light, partially interrupted band on each side of body from eye to base of hind leg. Ventral area white to yellow. Male with well developed helmet on head and a dorsal comb.

Distribution and habit: Southern Central America to northern South America, from northeastern Colombia through Panama to Costa Rica. Tropical rain forest. On trees or in bushes along the edges of water.

Care: Food: Insects, chicks, pink mice, fish, "moving" strips of beef. Should be kept in a semi-moist terrarium with an air temperature from 24 to 30° C during the day and 20 to 24° C at night; should be sprayed with water several times during the day. Localized heating with a radiator and regular UV radiation is suggested. Leafy soil is good as a substrate. Climbing facilities (branches, shrubs) must be supplied. Needs a large water dish for bathing.

Behavior and breeding: These tree-dwellers may even include fish in their natural diet. They catch fish by waiting motionless at the edge of the water, focusing on a fish passing by, then jumping on top of it, diving and grabbing it with their jaws. With the fish in their mouth they will then swim back ashore, smash their prey on the ground repeatedly to stun it, and then swallow it head-first. If in danger, the basilisk is also able to raise its body on to its hind legs and run away at great speed, just like some of the Australian agamids. They can achieve speeds of up to 12 km/hr in this "two-legged" running position. This can even take them across the surface of water without becoming submerged because their toes are

249

enlarged with skin folds.

The males usually display a pronounced territorial behavior. Therefore, in small enclosures such as terraria aggression among males is quite frequent. It is thus prudent to keep only one male with one or more females.

In nature the reproductive cycle of *B. basiliscus* is seasonally correlated. However, in captivity this quickly becomes modified so that these animals will then breed virtually at any month throughout the year. The terrarium must include warm, moist sites with leafy soil at least 25 cm deep. There the female digs alternately with the forelegs a hole about 11 cm deep and 9.1 cm in diameter, into which she lays 7 to 18 eggs in succession at 5-minute intervals. The female then turns around in the pit, examines the substrate with her snout, and closes the hole over with alternate shovelling with fore and hind legs. Finally, the surface of the site is levelled and further compacted with the snout. The female tends to remain for some time at this site, as if guarding the eggs, until the surface layer has actually taken on the correct appearance of the proper dampness level. Only then will the female leave and thereafter pay no more attention to the clutch. For more accurate monitoring and control it is advisable to carefully remove the eggs (21.5 to 25 mm x 11 to 14 mm in size), mark the upper sides, and place them in damp peat moss or a mixture of peat moss and sand inside a plastic container that in turn is kept inside a Type I brood container. At an incubation temperature of 26 to 35° C (according to habitat observations and measurements inside of eggs, the optimal incubation temperature must be around 30.5° C with minor variations) the young will hatch in 68 to 97 days. At birth they are about 10 to 13.3 cm long.

The newly hatched young should be transferred into a moist terrarium with a constant temperature of 27° C. On the second day after hatching (sometimes it may take up to 6 days) the young begin to hunt for small insects. Successful rearing is very much dependent upon a vitamin-rich, varied diet and regular UV radiation. At the age of 5 months the first fighting among males begins to occur. These fights will intensify as time goes on, so more compatible sub-groups should be formed in separate accommodations. This species becomes sexually mature at an age of 1½ years.

The following species can be kept and bred under identical conditions:

Basiliscus plumifrons (Iguanidae)—Green Basilisk
Maximum size: 70 cm.
Distribution: Central America.
Habitat: Tropical rain forest.
Clutch size: 11.
Incubation period and hatchling size: 60 to 66 days at 30 to 32° C; 13.3 cm.

Basiliscus vittatus (Iguanidae)—Banded Basilisk
Maximum size: 72 cm.
Distribution: Central America.
Habitat: Tropical rain forest, close to water.
Clutch size: 18.
Incubation period and hatchling size: 75 to 150 days at 25 to 35° C; 13 cm.

Corytophanes cristatus (Iguanidae)—Helmeted Iguana
Maximum size: 35 cm.
Distribution: Central America to northern South America.
Habitat: Tropical rain forest.

Calotes cristellatus (Agamidae)—Bloodsucker
Maximum size: 57 cm.
Distribution: Indonesia, Philippines, New Guinea.
Habitat: Tropical rain forest.
Clutch size: 2.
Incubation period and hatchling size: 56-65 days at 25-32° C; 13 cm.

Lyriocephalus scutatus (Agamidae)—Ceylonese Ballnose
Maximum size: 35 cm.
Distribution: Sri Lanka.
Habitat: Open tropical forest.
Clutch size: 14.
Incubation period and hatchling size: 141-146 days at 23 to 26° C; 6 cm.

Otocryptis wiegmanni (Agamidae)
Maximum size: 70 cm.
Distribution: Sri Lanka.

Habitat: Tropical rain forest.
Clutch size: 4.
Incubation period: 82 days at 23 to 25° C.

Chalcides bedriagai—Spanish Barrel Skink
(Family Scincidae)

Description: Size to 16 cm. Dorsum grayish brown to bronze with or without dark spots. A light lateral band present. Ventral area yellowish. Legs short. Small ear opening.

Distribution and habitat: Iberian Peninsula. Dry regions with loose substrate and plant growth.

Care: Food: Insects, spiders. Should be kept in a dry terrarium at air temperatures ranging from 25 to 35° C during the day and 18 to 22° C at night; supply localized heating with a heating coil and radiator plus regular UV radiation. As the substrate, use a layer of sand at least 8 cm deep, one half damp, the other dry. The terrarium should contain rock structures. Plants can include grasses, agaves, and similar items. This species can be over-wintered at 10 to 16° C, but this is not a mandatory prerequisite for breeding.

Behavior and breeding: This ground-dwelling species usually emerges from its sand or clay burrows in late morning and early afternoon to absorb the warm rays of the sun (supplied by a lamp in a terrarium). However, as soon as these skinks perceive ground vibrations they burrow—within seconds— into the substrate head-first with wriggling body movements. If this occurs in a dry, sandy substrate, the sand forms little wavy lines caused by the sand filling the burrowing track of the animal moving along below the surface.

Mating is not dependent upon the season; in fact, it can occur throughout the year. The male follows the female for several hours, repeatedly licking her cloacal region and trembling his body. If he succeeds in placing a mating bite laterally in her neck region, copulation follows quickly. To do that the male twists his tail below that of the female and grasps with one hind leg the base of the female's tail.

Spanish barrel skinks are live-bearers. After a pregnancy lasting 2 to 3 months, the female gives birth to 2 to 6 young of an average length of 5.4 cm. In order to avoid any possibility of cannibalism, the young are immediately transferred into a separate dry terrarium. The diet should consist of newly hatched crickets, small mealworms, small spiders,and similar insects. Regular UV radiation together with a varied vitamin-rich diet are prerequisites for breeding and rearing this species.

The following skinks can be kept and bred under the same conditions:

Chalcides chalcides
Maximum size: 48 cm.
Distribution: Iberia, southern France, Italy, northern Africa.
Habitat: Fallow regions, paddocks, escarpments with dense grass cover.
Litter size: 13.
Gestation period and juvenile size: 3 months; 9.5 cm.
Remarks: Sexual maturity in 2 to 3 years.

Chalcides ocellatus—Eyed Skink
Maximum size: 30 cm.
Distribution: Greece, Italy, Mediterranean Islands, North Africa, Arabia, Pakistan.
Habitat: Semi-dry regions with plant cover.
Litter size: 20.
Gestation period: 3 months.
Remarks: Should be kept in pairs only.

Chalcides viridanus
Maximum size: 25 cm.
Distribution: Tenerife, Gomera, Hierro (Canary Islands).
Habitat: Grasslands.
Litter size: 2.
Gestation period and juvenile size: 3 months; 8.2 cm.
Remarks: Sexual maturity in 2 to 3 years; should be kept in pairs only.

Ablepharus kitaibelli
Maximum size: 12 cm.
Distribution: Europe, Balcans, Asia Minor to Armenia, Syria, Israel, Sinai Peninsula.
Habitat: Deciduous forests to grasslands.
Clutch size: 4.
Incubation period and hatchling size: 2 months' incubation at

19-25° C; 3.3 cm.
Remarks: Egg-laying; sexual maturity in 2 years.

Chamaeleo chamaeleon—Common Chameleon
(Family Chamaeleontidae)

Description: Size to 30 cm but usually remains smaller. Base color green with two lateral rows of white or brown spots; however, coloration highly variable. White abdominal median band. Top of head roof-shaped with elevated helmet or casque. Low dorsal crest. Male with enlarged cloacal region.

Distribution and habitat: Iberian Peninsula (far south only) and some areas in the central and eastern Mediterranean region; also southwestern Asia and North Africa; introduced onto Grand Canary Island. Sun- and wind-exposed dunes in close proximity to coast with relatively dense brushland (*Lygos monosperma, Tamarix africana, Phragmites communis,* pines and eucalyptus forests with dense brush cover) as well as in desert regions with sparse vegetation.

Care: Food: Insects, spiders. Should be kept in a dry terrarium, greenhouse, or plant window at temperatures from 23 to 35° C and 16 to 22° C at night. Must be sprayed with water daily. Needs localized heating with radiator plus regular UV radiation. Adequate ventilation (one side glass, three sides wire mesh) is absolutely essential. The substrate should be fine sand, but a damp section 20 cm deep is essential for egg-laying. Climbing branches are necessary, and small lemon or orange trees or similar species as plants are also good. An automatic drop-waterer or manual watering with a pipette should be available. Can be over-wintered at 10 to 15° C

Behavior and breeding: Our field studies of this species in southwestern Spain and southern Portugal have shown that the common chameleon is solitary except during the mating season. Therefore, in captivity it should really be kept individually; males and females are brought together for mating purposes only. Alternatively, if a very large terrarium—or better yet a greenhouse or well-planted plant window—is available to give the animals the opportunity to establish territories without maintaining sight contact, more animals can be kept together in the same enclosure.

Territorial bonding in males is somewhat less pronounced only during the breeding season. This falls in the months of August and September on the Iberian Peninsula. The search for a partner can sometimes extend over substantial distances, in part even over sandy areas without plant cover (the females usually remain within their own area during this period). Due to the extensive movements of males during this period there is an increasing amount of confrontation with variable results, depending upon the sex, age, and motivation of the individual specimens involved.

If two males meet, they initially try to intimidate each other with threatening displays. This is accompanied by a color change to a uniformly green body color with black spots and bands; they also enlarge their body profile through lateral compression and spread their gular fold (throat pouch) with the help of the hyoid bone. The tail is curled up and the body profile is turned toward the opponent. The smaller of the two or the less motivated to fight then changes its color to dark grayish brown (the submissive coloration), makes himself as thin as possible, and moves away, always looking at the opponent until sight contact is finally lost. In the wild the dominant animal does not follow the defeated one, but in view of the spatial restrictions in a terrarium this does not hold true in captivity. So, outside the breeding season the animals should be kept individually in order to avoid mortalities.

What happens if both opponents are equally motivated? While going through their threatening displays both will move toward each other. When facing each other there are several twitching head movements to both sides, the tail is rolled out and back in a few times, and the opponents continue to move toward each other. They rise up on to their hind legs and fold the angled front legs together. With lurching forward and back body thrusts against the head and the base of the tail of the opponent, they try to push each other off the branch. Unless one of the rivals actually falls off the branch and then flees, displaying the submissive coloration, the ritualistic character of the fight becomes more and more lost as time goes on. Eventually one animal will firmly bite into the neck or back region of the other until that one displays the submissive coloration and flees. The winner does not follow the defeated animal.

If a male strays into the territory of a female he will display the threatening signals and approach the female. Her behavior then signals whether or not she is ready to mate. A female not yet ready to mate also will signal the threat display. Within seconds her coloration changes to black with many green and yellow dots. She also spreads her gular fold, inflates her body, and turns her broad side toward the male. Then she moves her body slightly from side to side several times. If the male continues to approach and attempts to mount her from behind she will rise up on her hind legs, turn around on her axis, open her mouth, and fend off the male with head butts, bites, or pushing with raised front legs, all accompanied by hissing sounds.

However, if she signals her mating readiness by remaining motionless and continuing to display the green body coloration, the male approaches slowly with accompanying twitching head movements from side to side. A few lurching head butts toward the back and base of tail of the female stimulate her to uncurl her tail and to lift it slightly at the base. The male then quickly grabs the back of the female and pulls himself on top of her. His hind legs reach for her tail base or hind leg. After some searching lateral movements the male brings his cloaca under the female's tail base. After the cloacal regions in both partners have become distinctly swollen the male inserts his hemipenis with accompanying up and down movements of his body. Copulation may last from 1 to 15 minutes. In contrast to many other species, the male does not bite into the neck of the female. The female signals her decreasing mating readiness by moving slightly forward.

About 50 to 60 days after copulation the female begins to search for a suitably damp and warm site on the sandy substrate to deposit her eggs. With her fore- and hind legs she will dig a pit about 15 to 20 cm deep into which she lays 15 to 50 eggs. The hole is then filled in again and carefully smoothed over. The eggs are 10-19 x 8-12.5 mm in size. They should be carefully uncovered, their upper side clearly marked, and then they should be transferred to a plastic dish with damp sand. The dish is placed inside a brood container of Type I. The incubation period is about 190 days at an air temperature of 27° C. The young chameleon will tear the tough egg shell with its egg tooth located on the premaxillary bone. The hatchlings are 47.5 to 51.5 mm long and weigh less than 1 gram. The young should be distributed into several dry terraria and be given a varied vitamin-enriched diet of small insects. Regular UV radiation is also essential.

Right: *Agama nupta fusca*. Photo by Dr. S. A. Minton. **Bottom:** *Acanthodactylus erythrurus*, the European fringe finger. Photo by E. Zimmermann.

Chamaeleo gracilis—Spur-heeled Chameleon
(Family Chamaeleontidae)

Description: Size to 34 cm. Green to yellow-brown base coloration, frequently with white or black spots or bands. Throat skin yellow-orange. Male with a spurred heel.

Distribution and habitat: Western, central, and eastern Africa. Gallery forest and regions around tropical rain forests; mango, orange, and acacia trees in the proximity of human habitation.

Care: Food: Insects, spiders. Should be kept in a semi-moist terrarium at air temperatures from 25 to 35° C during the day and 18 to 22° C during the night. Localized heating with a radiator and regular UV light is best. Adequate ventilation is absolutely essential (one side glass, three sides wire mesh). The substrate mixture of leafy soil and sand should include a damp area for egg deposition at least 10 cm deep. There should be adequate climbing facilities; plants such as orange, mango, and coffee trees or similar species are good. The terrarium should be sprayed with water in the morning. Use an automatic drop-waterer or offer water manually with a pipette.

Behavior and breeding: When in danger this species displays an avoidance behavior that is characteristic for many representatives of the family Chamaeleontidae. If a potential enemy approaches, the chamaeleon will make itself as thin as possible. Muscular action enables it to flatten the body laterally; the animal then turns around and actually hides (e.g., behind a branch) and remains motionless to rely on its camouflage coloration. If an enemy attacks after all, the chameleon inflates its body and the coloration changes dramatically to green with black spots and cross bands. The bright orange throat sac is extruded as a threatening gesture, and the body is raised on the hind legs. With an open mouth, hissing, and front legs raised in a defensive posture, the chameleon tries to scare off the enemy. We have actually seen this sort of behavior in the wild in West Africa when chameleons were approached by humans. However, in captivity the chameleon will quickly become used to people.

The solitary nature of these animals should be catered for in captivity. For instance, in a well-planted greenhouse confrontation with other chameleons is largely avoided. If indeed contact is made they will immediately display threatening behavior. The green body coloration with numerous black dots appears, the bright orange-colored throat skin is displayed and the body profile, enlarged by flattening the body laterally, is turned toward the opponent. This is an attempt to intimidate the opponent and so persuade him to leave. Usually one of the two will sooner or later turn away, making himself as thin as possible and taking on a dark coloration. This individual then flees or remains motionless with his body pressed firmly against a branch. This submissive gesture pacifies the opponent and lessens the threat of aggressive action.

However, if both animals repeat these threatening gestures and try to place themselves in an elevated, strategically better position, a fight ensues that can have fatal consequences. For a fight both animals rise on their hind legs and with their head and forelegs raised butt each other, trying to bite into each other's throat region. When one of the opponents tries to flee and so climbs down from the tree, the other one will follow to keep on biting into the throat and body region until bleeding wounds occur. Therefore, it is better to keep these animals individually or in terraria that are sufficiently large and adequately decorated so that there is no visual contact among the animals.

Little is known about the reproductive behavior of this species. The female digs pits about 8.5 cm deep into moist, warm areas of the substrate. Up to 45 parchment-like eggs (14 mm long and 10 mm in diameter) are deposited in such a pit. Subsequently the pit is closed again and the surface is smoothed over so that there is virtually no evidence of the egg pit. The incubation period lasts 219 days. For details about rearing the young please refer to *Chamaeleo jacksonii*.

Chamaeleo hoehneli
(Family Chamaeleontidae)
Description: Size to 20 cm. Base coloration blue-green to brown with yellow or dark green lateral spots or bands. Ventral median line white to gray.
Distribution and habitat: Eastern Uganda, Kenya. Bush zones, outer forest regions, and bamboo zones of highlands up to 3000 m.
Care: Food: Insects. Should be kept in a semi-moist terrarium with air temperatures from 20 to 25° C during the day and 13 to 17° C at night. The terrarium must be sprayed

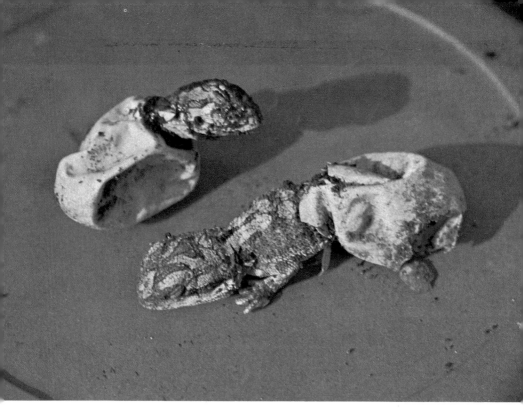

Amphibolurus barbatus, bearded dragons. **Top:** Hatchlings. Photo by Dr. S. A. Minton. **Bottom:** In captivity this species develops a hierarchy. Photo by E. Zimmermann.

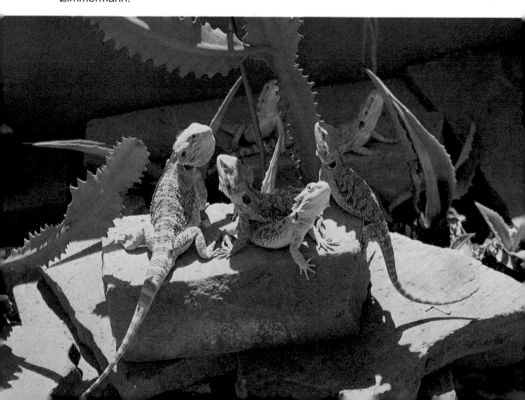

Mating and birth in the slowworm, *Anguis fragilis*. Photos by D. Poisson, courtesy Dr. D. Terver, Nancy Aquarium, France. **1)** Copulation. **2)** Detail of mating bite. **3)** Birth.

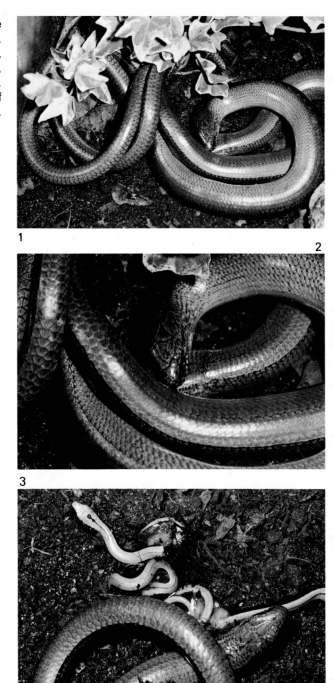

1

2

3

daily with water. Provide localized heating with some type of radiator and also regular UV lights. Adequate ventilation is absolutely essential (one side glass and three sides wire mesh works well). Use a sand and peat moss mixture as a substrate and include climbing branches. Plants that can be used include coffee, lemon, and orange trees or similar species.

Behavior and breeding: Despite the relatively small size of this species, a large terrarium is needed if more than one specimen is to be kept. As in many other chameleon species, the males establish territories that are vigorously defended against all other males. If a territory holder sees another male, he will attempt to intimidate the intruder with threat displays. For that he flares his throat skin with the aid of the hyoid bone, flattens his body laterally, rolls up his tail, and moves toward the intruder, accompanied by jerky up and down motions of the head. If the intruder does not move away, display a dark submissive coloration, or drop to the ground, a fight begins. Such a fight can last up to 5 minutes, and in a small enclosure this can have serious consequences. The opponents encircle each other, continually threatening each other and always maintaining the body profile toward each other. With their forelegs folded they rise on to their hind legs, opening the mouth wide and trying to bite the enemy's throat sac, dorsal comb, base of tail, or any of the other extremities. In contrast to sexually mature males, females are tolerated in a male's territory.

As in most other representatives of the family Chamaeleontidae, mating foreplay in this species is initiated by the male. Nodding his head and displaying his gular fold, the male approaches the female. About 20 cm in front of her he takes on a light green coloration. A female that is already pregnant will signal its mating unwillingness by changing color to a dark brown with lighter lateral bands, swaying sideways, and threatening with lateral flattening. Such a female will also rise on its hind legs, opening its mouth in a threatening manner.

However, females willing to mate will respond by nodding their head slightly and (usually) remaining motionless. The male will approach the female from behind. He holds on to the back of the female and winds his tail under hers. This is followed by copulation, which lasts about 10 minutes.

Chamaeleo hoehneli is live-bearing. About 6 months after

copulation the female becomes increasingly unsettled and constantly walks about in the terrarium. Soon thereafter— usually during the late morning hours—the birth of 8 to 21 young takes place. They are 46 to 53 mm long. At first they remain inside their embryonic membranes with the extremities close to the body and the tail rolled up. They tend to stick to branches, leaves, and other objects. The sudden daylight stimulus causes the young to stretch inside the membranes shortly after birth. They make twitching movements inside the membranes that cause them to burst open, and the young begin to breathe for the first time.

Immediately after hatching the young are still light and colorless, but within 10 minutes they start to take on their contrast-rich juvenile coloration, which is dark brown to green with two rows of light triangular lateral spots. During the following few hours the young try to disperse as far as possible, presumably so that they do not get eaten by the female. Even at that early age two siblings will already threaten each other should they accidentally meet.

The juveniles are transferred into a semi-moist terrarium (possibly a wire-mesh container with a glass front—ventilation!) and are kept at low temperatures because even air temperatures in excess of 25° C have proved to be detrimental. Within a few hours after birth the young will already shoot at *Drosophila*, small spiders, grasshoppers, and caterpillars with their relatively long tongues. A vitamin-rich, varied diet with occasional calcium supplements and regular UV radiation are essential for rearing the young of this species successfully. Sexual maturity is reached in 9½ months.

Chamaeleo jacksonii—Three-horned Chameleon
(Family Chamaeleontidae)
Description: Size to 32 cm. Body color green to yellow-brown, sometimes with white or brown spots. Serrated dorsal comb present. Three long horns on top of the snout in males; in females the horns are usually smaller or only one horn or conical scales may be present.
Distribution and habitat: Eastern Africa (Tanzania, Kenya). Highland forest areas up to 3000 m elevation.
Care: Food: Insects, spiders, pill bugs. Should be kept in a semi-moist terrarium at air temperatures from 22 to 26° C during the day and 15 to 18° C at night. Provide localized

Top: *Anolis carolinensis*, the green anole. Photo by J. Dommers. **Bottom:** *Anolis garmani*, the Jamaican anole. Photo by H. Zimmermann.

Right: *Chalcides viridanus*, a barrel skink, with young. Photo by H. Zimmermann. **Bottom:** *Chamaeleo chamaeleon*, the common chameleon. Photo by H. Zimmermann.

heating with a radiator and regular UV radiation. Adequate ventilation is absolutely essential; if possible, use a cage with one side glass and three sides wire mesh. A leafy soil and peat moss mixture is good as a substrate. Climbing branches are necessary, and decorative planting with small orange or coffee trees or similar species is good. The terrarium should be sprayed with water in the morning. Will take water from an automatic drop dispenser or from a pipette.

Behavior and breeding: When a three-horned chameleon goes hunting for prey, it initially scrutinizes a large area in its vicinity using both eyes (which can move independent of each other). Once suitable prey has been spotted, both eyes will be fixed upon it. Then the animal moves toward the prey in slow-motion, points the head directly at the prey, and suddenly explosively projects its tongue toward the prey, grasping the insect with it and returning the tongue and prey back into the mouth, where the prey is then eaten. Yet a hunt does not always work this way. Young chameleons are rather clumsy and tend to miss their aim, so they first have to learn to estimate the distance accurately. However, even the more competent catching methods among the experienced animals are still dependent upon many different factors, such as the motivation of the animal, the type and movement of the prey, and the air temperature and light intensity.

The males are—as in the case of many other species of this genus—rather incompatible when together. If two of them meet—which is nearly always the case in a terrarium—they initially attempt to establish a dominant relationship by threatening each other. One tries to intimidate the other with a rapid color change to bright green, displaying the throat pouch, raising the body, flattening the body laterally, and finally raising the forelegs and directing them toward the opponent. Often this works and one of them takes on the submissive coloration, black with narrow green lateral bands. Alternatively, the submissive male tries to interrupt sight contact with the dominant animal by hiding, moving away, dropping to the ground, flight, or akinesis ("freezing"). The "loser" will frequently hide in a lower corner of the terrarium and is thus prevented from feeding. Therefore, it is recommended that males be separated.

However, if both animals are equally motivated they further enforce their threat displays with open mouths, bodies

266

leaning toward each other in parallel positions, and raising the body together with repeated forward and back movements. Then a fight will ensue that can last from 10 to 94 minutes. Such a fight contains ritualistic as well as destructive (causing physical damage) elements. Both opponents attack each other with the head lowered, the preorbital and nasal horns pointing at each other, in an attempt for one to dislodge the other. As the fight drags on the ritualistic elements gradually become lost and one opponent will try to bite the other in the neck or throat. If one of them succeeds in pushing the other one off the branch or if one of them assumes submissive coloration and begins to retreat, the dominant animal will follow and continue to attack as long as there is visual contact. Therefore, it is advisable to keep males individually or together with only one female in a large terrarium.

Females also tend to be somewhat incompatible at times, threatening each other. However, this involves only inflating the body and lateral body movements, together with a minor coloration change. Actual fighting among females is rather rare.

If a male is introduced into a terrarium with a female, he will attempt to impress her with a threat display. However, the female invariably reacts with a threat rejection and keeps her distance. She spreads her gular fold and raises the body on its hind legs. She inflates her body, opening her mouth in a threatening manner, while the body sways several times from one side to the other. If the male continues to approach and attempts to mount the female from behind, she turns around and tries to fend off the male by changing coloration to black with irregular green lateral spots and by raising the forelegs.

If the female is willing to mate she merely displays a weak threatening behavior at some distance or remains totally motionless. The male recognizes the willingness by her behavior and approaches, his head twitching repeatedly to left and right accompanied by similar eye movements. The female then changes to a light ocher color and slowly unrolls her tail. The male then makes body contact and reaches around her flanks or to the base of her tail. She in turn bends her back down and at the same time elevates her tail slightly. The male then reaches for her neck and pulls himself up onto her

Chamaeleo hoehneli. Photo by B. Kahl.

Chamaeleo jacksonii, the three-horned chameleon. **Top:** One-day-old young. Photo by J. Bridges. **Bottom:** Adult male. Photo by E. Zimmermann.

back. After having stabilized his position, he maneuvers his cloacal slit below the female's tail. Following a distinct cloacal swelling in both animals, the male inserts his hemipenis. This is followed by contractions in the male's cloacal region in rhythmic sequences of about 2 seconds' duration. The entire mating procedure lasts about 13 minutes. After about 8 to 9 minutes the female signals the end of her mating willingness by swaying her body slightly, which increases during the following 1 to 2 minutes. At the same time she changes her coloration toward darker tones and usually also moves a few steps forward. This induces the male to withdraw the hemipenis, and both partners move away (the male quite often displays some more "head twitching"). The female retains her mating willingness for a maximum of 11 days, a period during which she copulates on 6 separate days. However, if a male attempts to copulate with her twice on the same day he is driven off.

During the following pregnancy period, which lasts on the average 192 days, the female has an increased demand for vitamin-enriched food. However, 4 weeks before giving birth her food intake declines. One to 2 days before the actual birth the female begins to pace restlessly through the terrarium.

Delivery usually takes place during the morning hours. There are no conspicuous outward signs of contractions. The female everts her cloaca about 2 or 3 times, and after another 40 seconds the first young is born. At birth it is still surrounded by a sticky, gelatinous egg membrane and rests motionless in its embryonic position inside the egg. The eyes will open only after contact with the substrate has been made, and after another 92 seconds the young begins to turn around. With a sudden stretching of the body and legs, the egg membrane bursts. The young pushes its head out, begins to breathe, and then crawls out of the membrane. This tears the umbilical cord. The characteristic juvenile markings, black with triangular white lateral spots, are present. In the meantime the female has moved on and the next young are born. She does not pay any attention where the eggs are dropped, so it can happen that they stick to twigs and branches or simply fall onto the ground. It is also possible that a particular egg membrane begins to dry very rapidly and prevents the young from freeing itself. To avoid such accidents we can help by opening the membrane cautiously with a pair of forceps and scissors. However, this procedure is rarely ever successful, the young invariably being simply too weak or deformed, so they usually die soon after assisted birth.

Duration of the entire birth process depends on the size of the litter; it can last from 32 to 225 minutes, during which up to 38 young can be born. 20 days after having given birth the female can copulate again.

At birth the young are about 5.5 cm long, with an average weight of 0.6 grams. They should be separated from their mother and raised in small groups maintained in well-ventilated semi-moist terraria. With an adequate diet consisting of vitamin-enriched *Drosophila*, small crickets, wax moth larvae, mealworms, moths, small grasshoppers, spiders, and similar foods, the young will double their weight in one month and their size in about 6 months. They should also be given regular UV radiation and the terrarium must be misted with water every morning. Sexual maturity is attained at an age of 9 or 10 months; at that stage they display a well-defined social behavior and it is advisable to keep them individually or in pairs only.

Coleonyx variegatus—Banded Gecko, Banded Clawed Gecko

(Family Gekkonidae)

Description: Size to 14 cm. Dorsum yellow with brown cross-bands. Movable eyelids with yellow margins. Toes without adhesive pads. Male with strong spine on each side of tail base.

Distribution and habitat: Southwestern USA. Desert regions; rocky crevices, under loose rocks, and in rodent burrows.

Care: Food: Insects, spiders. Should be kept in a dry terrarium at air temperatures from 25 to 35° C during the day and 18 to 22° C at night. The terrarium must be sprayed with water in the evening. The substrate should consist of sand (⅓ damp section, ⅔ dry area) with rock structures as decoration. Plants (succulents) can be omitted. Provide a small water bowl.

Behavior and breeding: This species is nocturnal and ground-dwelling. It pursues its prey with the body elevated on all four legs. Once prey has been spotted the tip of the tail quivers in excitement. The gecko moves slowly toward the prey,

Top: *Cnemidophorus lemniscatus*, a tropical racerunner. Photo by Ken Lucas at Steinhart Aquarium. **Bottom:** *Coleonyx variegatus*, the banded gecko. Photo by B. Kahl.

Top: *Corytophanes cristatus*, the helmeted iguana. Photo by A. Norman. **Bottom:** *Cordylus warreni depressus*, a girdle-tailed lizard or sun gazer. Photo by A. Norman.

stops briefly to observe it, and then thrusts forward to grab it with the jaws and swallow it.

If threatened by an enemy such as a snake and there is no possibility of escape, this species, like many other geckos, has a novel and characteristic defensive mechanism: It raises its conspicuously colored yellow and black banded tail into the air and begins to move it with undulating motions. This distracts the enemy's attention to the tip of the tail. If the enemy jumps at it the gecko severs its tail along a predetermined fracture line and the snake bites into what it has perceived as prey and the gecko escapes. As in many other lizards, the tail of this gecko will eventually regenerate.

The natural reproductive period extends from April to August but in captivity can vary. During the courtship the male approaches the female with his head and body close to the ground and the tail moving excitedly. With its head and snout the male pushes against the sides and cloacal region of the female, sometimes also licking these areas. Then he bites at her tail base, hind legs, or other posterior area. Suddenly the male slides onto the back of the female, always biting a section further forward to secure a new hold, until he can finally place the characteristic mating (neck) bite. At the same time his hind legs maneuver for a suitable mating position, and in doing so he moves repeatedly with his cloacal region over the base of the female's tail. A female willing to mate will then raise her tail slightly so that the male can bring his tail under hers and anchor his tail spur at her cloacal region. This causes the cloacal lips of the female to be pulled back, and the male can thus insert his hemipenis.

A few weeks after copulation the female deposits two parchment-shelled eggs at a damp site on the substrate. Their upper side should be marked and the eggs then transferred into a plastic container filled with damp, sterilized sand, which is then placed inside a Type I brood container. The young will hatch in 42 to 58 days at a temperature of 22 to 30° C. At birth they are 5.4 to 5.7 cm long. They should be reared individually, preferably in plastic refrigerator containers (use a wire mesh cover for ventilation; decoration and care same as for adults). The food should consist of small crickets, larvae of wax moths or flour moths, or similar insects. With a varied, vitamin- and mineral-enriched diet this species reaches sexual maturity in about 10 months.

Cordylus cordylus—Girdle-tailed Lizard, Sun Gazer
(Family Cordylidae)

Description: Dorsum dark brown to greenish yellow with regular thorn-like scales. Tail scales strongly keeled, in a whorl-like arrangement. Male with well-defined thigh spurs.

Distribution and habitat: Ethiopia to South Africa. Rocky regions.

Care: Food: Insects. Should be kept in a dry terrarium at air temperatures from 23 to 28° C during the day and 13 to 18° C at night. Use localized heating with a radiator and regular UV radiation. The terrarium must be sprayed with water morning and evening. Use sand as a substrate and flat rocks and rocky structures as decoration. Succulents and similar species of plants can be added. A small water dish should be added.

Behavior and breeding: Cordylus cordylus (as well as *C. cataphractus*) displays an interesting defense behavior that, however, is quickly lost in captivity. This species will bite into its own tail to form a circle with its body and so present a thorn-armored surface to any potential enemy. The rather weakly armored abdominal region is thus protected.

Little is known about the reproductive behavior of this ground-dwelling species that is active primarily during the day. However, it has been bred in captivity. Like all girdle-tailed lizards, this species is a live-bearer.

A female can give birth to from 2 to 4 young that are about 7 cm long. The young should be accommodated separately in a dry terrarium. The diet includes crickets, wax moth larvae, mealworms, grasshoppers, and other small insects. This species requires a lot of sunlight, therefore regular UV radiation is necessary, particularly for rearing the young.

> *The same conditions for keeping and breeding also apply to: Cordylus cataphractus (20 cm; southwestern Africa; savannah and desert) with 1 to 2 young per litter; and Cordylus giganteus (40 cm; South Africa; savannah and desert) with 2 to 4 young per litter.*

Egernia cunninghami—Cunningham's Skink
(Family Scincidae)

Description: Size to 36 cm. Dorsum grayish brown to black

Top: *Diplodactylus vittatus.* Photo by Dr. S. A. Minton. **Bottom:** *Crotaphytus collaris*, the collared lizard. Photo by Ken Lucas at Steinhart Aquarium.

Right: *Eublepharis macularius*, the leopard gecko. Photo by B. Kahl.
Bottom: *Eumeces schneideri*, the spotted or Berber skink. Photo by H. Zimmermann.

with white dots or spots. Ventral area white to gray, frequently with dark spots. Thorn-like extended scales dorsally. *Distribution and habitat:* Southeastern Australia. Rocky areas. *Care:* Food: Insects, newly born mice, snails, fish, pieces of beef, fruit, vegetables, clover, dandelion. Should be kept in a dry terrarium at air temperatures from 25 to 37° C during the day and 18 to 22° C at night. Provide localized heating with a radiator and floor heating, plus regular UV radiation. Use sand as a substrate. Decorations can include rocky structures and climbing facilities, as well as succulents, but plants can be omitted if desired. A small water bowl should be available.

Behavior and breeding: This species—like most skinks—is a ground-dweller. It always stays close to crevices and is able to flee quickly into the rocks when danger approaches. By inflating the body the skink becomes firmly wedged in crevices and points its spiny tail toward the pursuer. If the tail is actually grabbed, it tends to break off at a predetermined fracture line and the skink can still escape.

Males of this territorial skink are rather aggressive toward each other in a terrarium. Often this leads to fights and serious bite wounds. Therefore, it is recommended that they be kept individually or in pairs only.

Egernia cunninghami is being bred regularly. During the courtship the male follows the female, encircles her with his gular pouch inflated, and tries to bite firmly anywhere along her body. When he finally succeeds in establishing his mating bite at the female's upper arm and effectively stops her from escaping, copulation follows shortly.

About 100 days later the female will give birth to 2 to 6 young during a period of 2 to 7 hours. If at birth they are still attached to the yolk sac via an umbilical cord, the female will cautiously pick up the yolk sac in her jaws, bite off the cord, and eat the yolk sac.

The young are about 12 cm long and weigh 8 grams. They have a distinct juvenile coloration, being jet black with glowing white spots. In order to assure better control and supervision, the young should be accommodated in separate dry terraria. A few hours after birth the young begin to lick at sweet fruit and chase after small insects. Sexual maturity is attained after about 3 to 6 years. Successful rearing requires a varied vitamin-rich diet and regular UV radiation.

Egernia stokesii—Gidgee Skink
(Family Scincidae)

Description: Size to 24 cm. Dorsum red to olive-brown with numerous lighter spots. Spiny scales particularly well developed along the depressed tail. Ventrally yellow-brown. Males tend to have slightly longer tails than females.

Distribution and habitat: Australia, primarily in the dry interior, the open, rocky plains.

Care: Food: Insects, fruit, vegetables, newly born mice, raw eggs, puddings, and similar items. Should be kept in a dry terrarium at air temperatures from 25 to 37° C during the day and 18 to 22° C at night. Provide localized heating with a radiator of some type as well as regular UV lighting. Sand is good as a substrate. Rocks and climbing branches make good decoration, as do succulents or similar species of plants, but plants can be omitted. A small water bowl should be available.

Behavior and breeding: These sun-loving skinks are mainly active during the day. When attacked they display a characteristic defensive behavior. If one of them is cornered and cannot escape, it first threatens the aggressor with a wide-open mouth. Then the skink displays its deep-blue tongue set against a background of red mucous tissue—a distinct warning signal. If the skink is actually picked up it will attempt to bite by suddenly flexing its body and will also hit the aggressor with the beating tail. The latter is indeed an effective weapon, and wounds to a human hand caused by the spinous tail usually take a long time to heal.

The senses of smell and taste seem to play a significant part in hunting for prey as well as in social behavior. If food is placed in the terrarium the animals will appear one after the other either from the sunning location or out of hiding. They will push their tube-like body with the short legs toward an insect, flicking their tongue several times, and then lunge at the prey, grabbing it in their jaws and swallowing it. Fruit and other sweet items are tested first with the tongue before being eaten.

Initially there is no aggression among members of a family group (we are keeping an adult pair and 5 young, born in the course of 3 years, together). However, once the young reach an age of about 3½ years (i.e., the age of sexual maturity) the behavior of one parent changes rather dramatically toward

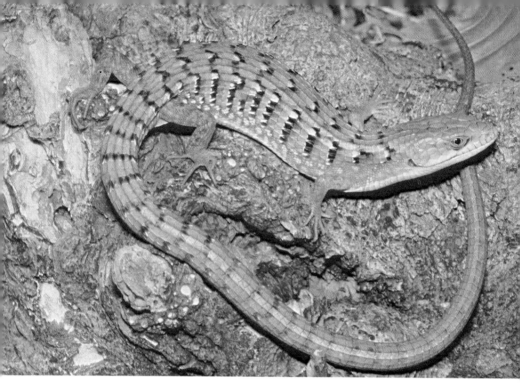

Top: *Gerrhonotus multicarinatus*, the southern alligator lizard. Photo by B. Baur.
Bottom: *Egernia stokesii*, the Gidgee skink. Photo by H. Zimmermann.

Top: *Gonatodes fuscus*, the yellow-headed gecko. Photo by J. T. Collins. **Bottom:** *Gekko gecko*, the tokay gecko. Photo by B. Kahl.

siblings of the same sex. The older *female* (and *not* the older male as in many other species) then begins to attack the younger females and tries to drive them out of the family group. Flicking its tongue, the old female approaches one of her daughters with lateral, twitching motions of head and tail and tries to bite into the spinous tail of the younger animal. The young female avoids the attack, turns around, flattens her tube-like body against the ground, and offers one flank to the older female; this offers little attack opportunity, yet the old female persists with her attacks. She approaches the young animal from behind and tries to bite it in the tail or a foot. If she succeeds in securing a firm bite into a foot, she will quickly toss the younger animal onto its back. Usually one of the tiny, thin toes becomes a casualty by being severely bitten or is even completely severed. Such fights can easily lead to the loss of several toes, so it is advisable to keep sexually mature females separated. If placed together with a male of similar age, copulation will take place quickly and further progeny is assured.

Thus in the now-reduced family group peace will quickly return and the old female once again becomes the alpha (dominant) animal. Since *E. stokesii* loves to sunbathe, these animals look for suitable sun-exposed sites. If a younger animal happens to occupy such a site and is approached by an older female, the latter will lick the youngster. Then the youngster's tail begins to move from side to side, and the legs are lifted alternately and in the same sequence. Such submissive behavior—reminiscent of the "leg kicking" ("treteln") among other lizards—tends to inhibit aggression from dominant animals.

The mating season for imported animals as well as for the F1 generation occurs during September and October. The male follows a female for several days (she tends to evade the male) and then licks her cloacal and tail region. If the female signals her willingness to mate by pressing her body flat to the ground and lifting her tail, the male applies the mating bite laterally to the neck of the female and holds on to her back and tail with his front and hind legs. He then maneuvers his tail under that of the female. Copulation (lasting about 14 minutes) occurs following insertion of the hemipenis and is accompanied initially by regular up and down motions of the tail.

282

This species is a live-bearer. After a pregnancy of about 3½ months the female gives birth to usually 1 or 2 young that are 8 cm long and weigh 6 grams. Soon after birth the young should be transferred to a separate dry terrarium. Even at such an early age they already display the same characteristic defensive behaviors—opening of mouth and biting—as their parents. A few hours after birth the young will begin to chase after small crickets, grasshoppers, and similar small insects. It is important that these growing youngsters be given a varied, vitamin-rich diet, together with calcium supplements and regular UV radiation (such as Osram Ultra Vitalux-type lighting at 1 meter distance, twice a week for 5 minutes each, or with UV-A fluorescent tubes 12 hours per day).

The follow species can be kept and bred under identical conditions:

Egernia inornata
Maximum size: 20 cm.
Distribution: Central Australia.
Habitat: Desert.
Litter size and juvenile length: 2-3; 5 cm.

Egernia striolata
Maximum size: 22 cm.
Distribution: Australia.
Habitat: Dry regions.
Litter size and juvenile length: 4; 10 cm.

Eublepharis macularius—Leopard Gecko
(Family Gekkonidae)
Description: Size to 30 cm. Dorsum light gray to yellow with bluish black spots. Ventral area yellowish white. Male with pre-anal pores. Movable eye lids.
Distribution and habitat: Eastern Iran to Pakistan. Rocky deserts and semi-arid grasslands, under rocks or in underground burrows.
Care: Food: Insects, spiders. Should be kept in a dry terrarium at air temperatures from 25 to 40° C (less at night). The terrarium should be misted with water in the morning. Use sand or fine gravel as a substrate and provide large rocks or roots as hiding places and a small water bowl.

Top: *Hemidactylus frenatus*, a house gecko. Photo by S. McKeown. **Bottom:** *Hemidactylus turcicus*, the Mediterranean gecko. Photo by E. Zimmermann.

Lacerta agilis, the sand lizard, courting pair, the male with green flanks. Photo by B. Kahl.

Behavior and breeding: In a terrarium leopard geckos tend to dig burrows under large roots or below rocks; the burrow is sufficiently large enough for the animal to stand up in it. This species is truly nocturnal.

These geckos stalk their prey with the body elevated on all four legs. Once an insect has been spotted they fix their sight on it and run slowly toward it with the tail slightly raised and moving from side to side and forming an S. They will stop a short distance from the prey then lurch at it, seizing it with their jaws and swallowing it.

Little is known about the reproductive behavior. The female selects a damp site on the substrate to deposit her eggs. There she digs a hole about 5 to 8 cm deep into which the soft-shelled eggs (30 x 20 mm in size) are laid. Subsequently the clutch is carefully covered over again with sand. The eggs should be uncovered and their upper side carefully marked. Then they are transferred to a plastic container with damp sand that is placed inside a Type I brood container. The eggs will hatch in 39 to 53 days at a temperature of 23 to 30° C. The young are 83 mm long and weigh 3.6 grams. Newly hatched specimens have interesting markings, brownish black bands against a yellow background. It is advisable to move the young into separate accommodations in a dry terrarium. They will soon feed on small crickets, grasshoppers, and the like.

Eumeces schneiderii—Spotted Skink, Berber Skink
(Family Scincidae)

Description: Size to 47 cm. Dorsum brown with white, yellow, or orange dots or patches. A yellow lateral band. Abdomen whitish yellow. Some herpetologists consider *Eumeces algeriensis* to be a subspecies of *E. schneiderii*, while others believe it to be a species in its own right.

Distribution and habitat: Northern Africa (from eastern Algeria) to central Asia (Tadshikistan, northwestern India). Semi-desert to dry agricultural areas, among bushes and under rocks.

Care: Food: Insects, spiders, snails, newly born mice, chopped meat. Should be kept in a dry terrarium with air temperatures from 25 to 35° C during the day and 18 to 22° C at night. Supply localized heating with a radiator and floor heating plus regular UV lighting. In the morning and evening mist the tank. Use sand at least 10 cm deep as a substrate. The decoration can include rock slabs, tiles, and pieces of bark as hiding places. Succulents may be added. Give a small water bowl.

Behavior and breeding: Spotted skinks live not only on the ground but also in shallow burrows under ground. As an adaptation to this mode of life the species has several scales in a comb-like arrangement in front of the ear drum that protect the ear to some extent against penetration of sand and dust.

Males are particularly aggressive during the mating season, when their fighting can inflict substantial injury, especially damage to the tail. If this should happen the injured animal must be removed immediately. During the courtship the male approaches a female, licks her cloacal region, and tries to secure a mating bite at her neck.

Five to 6 weeks after copulation the female moves restlessly through the terrarium looking for a suitable place to deposit her eggs. From 3 to 20 oval white eggs are eventually deposited among rock slabs or in slightly damp sand. The female provides brood care by wrapping her body around the clutch and remains there for 8 to 9 weeks until the eggs hatch.

Eumeces obsoletus (Great Plains Skink; 35 cm; southern USA, Mexico; rocky areas in the proximity of water) can be bred under identical conditions. The female guards the clutch of up to 17 eggs, which hatch in about 1½ months. Sexual maturity is attained after 2 years. *Eumeces fasciatus (Five-lined Skink; 20.5 cm; southern USA; damp meadows with leaves and logs),* female guards clutch (up to 15 eggs) for 1½ months until young hatch; should be kept in a semi-moist terrarium.

Hemidactylus flaviviridis—Indian House Gecko
(Family Gekkonidae)

Description: Size to 16 cm. Dorsum light yellow to olive brown with undulating dark bands. Ventral area yellowish. Males with pre-anal and femoral pores.

Distribution: Iraq to Bengal. On brick and stone walls, in crevices and under loose tree bark.

Care: Food: Insects, spiders. Should be kept in a dry terrar-

Left: *Lacerta lepida*, the pearl or eyed lizard. Photo by B. Kahl. **Bottom:** *Lepidodactylus lugubris*, the mournful gecko. Photo by S. McKeown.

Right: *Lygodactylus picturus*, the dwarf gecko. Photo by H. Zimmermann. **Bottom:** *Mabuya quinquetaeniata*, the African blue-tailed skink. Photo by B. Kahl.

ium at temperatures from 23 to 30° C during the day and at 18 to 22° C at night. Provide localized heating with a radiator. Should be sprayed in the morning. Use sand as a substrate. Decorations can include climbing branches and possibly a cork bark back wall. Succulents can be planted. A small water bowl should be provided.

Behavior and breeding: This is essentially a nocturnal species, with males establishing territories that they defend against other males. If one male trespasses into the territory of another a fierce fight ensues. The males chase and follow each other, emitting combat sounds and attempting to bite each other. The strongest and most persistent male invariably ends up being the winner. Usually this male then also goes on to court a female. It encircles a potential partner, approaches her while nodding his head, and touches her flank with his snout. He then turns away again and renews his encircling maneuver. After a courtship display of about 10 minutes he climbs slowly over the back of the female and moves his tail in a sinusoidal (S-shape) motion. At the same time his hind legs move searchingly for a suitable mating position. The male then applies the mating bite to her neck, slides up against her side, and moves his tail under hers. Following cloacal contact, copulation occurs during the following 4 to 10 minutes.

About 2 weeks later the female deposits 2 eggs under rocks or pieces of cork or even into the sand substrate. The eggs should be removed and placed in a Type II brood container. The young will hatch after 62 days at an air temperature of 26° C. For rearing details see *Hemidactylus turcicus.*

The following *Hemidactylus* species can be kept and bred under the same conditions:

Hemidactylus brooki
Maximum size: 14.5 cm.
Distribution: West Africa.
Habitat: Buildings, trees, termite nests.
Clutch size: 2.
Incubation period and hatchling size: 47-55 days at 32° C; 4.6 to 5 cm.

Hemidactylus frenatus
Maximum size: 12 cm.

Distribution: Southern and eastern Africa to southeastern Asia to Australia, Oceania, and Mexico (travels widely with man).
Habitat: Buildings, walls, rocky areas.
Clutch size: 2.
Incubation period and hatchling size: 56 days at 30° C; 3 cm.

Hemidactylus mabouia
Maximum size: 12 cm.
Distribution: Central, southern, and eastern Africa, Madagascar; introduced into South America, Antilles, Central America.
Habitat: Buildings, walls, rocky areas.
Clutch size: 2.
Incubation period and hatchling size: 80 days at 25 to 30° C; 2.5 cm.

Hemidactylus turcicus—Mediterranean Gecko
(Family Gekkonidae)
Description: Size to 12 cm. Dorsum reddish brown to grayish black with irregular dark patches. Belly white. Tail usually banded in dark to light brown. Undersides of fingers and toes with two longitudinal rows of oval-shaped adhesive lamellae. Male with enlarged cloacal region.
Distribution and habitat: Island and coastal areas of the Mediterranean Region, North Africa, and southwestern Asia. Introduced into North and Central America. On walls and cliff faces, in caves, on tree trunks, and among agave plants in primarily dry areas.
Care: Food: Insects, spiders. Should be kept in a dry terrarium at air temperatures from 25 to 35° C during the day and 16 to 22° C at night. Provide localized heating with a radiator and spray daily. Sand or a sand and peat moss mixture makes a good substrate. Decorate with rock structures, climbing branches, and a cork bark back wall. Succulents or similar plants can be given but also may be omitted. Provide a small water bowl.
Behavior and breeding: Although this species sometimes emerges from its hiding place even during daylight hours to soak up some sun on a wall or other exposed area, it is primarily active during dusk and at night. Prey is caught at night, often near a street light or some other light source that attracts insects.

The males display a pronounced territoriality, particularly

Top: *Oedura monilis*, the ocellated velvet gecko. Photo by E. Zimmermann. **Bottom:** *Ophisaurus apodus*, the scheltopusik or European glass lizard. Photo by B. Kahl.

Phelsuma guentheri, a day gecko. Photos by Coffey and Bloxam, courtesy Dr. D. Terver, Nancy Aquarium, France. **Top:** Juvenile. **Bottom:** Adult female.

during the breeding season. We were able to observe this in our dry terrarium (60 x 40 x 30 cm). If one male enters the territory of another male the resident male emits some muffled "go-go-go" sounds to proclaim his territorial occupancy. If the trespasser does not move away, the resident male will then try to drive the invader away. He raises his body on all four legs, inflates his throat, and signals his readiness to attack with a sequence of rapid "woh-woh-woh" sounds. He focuses his eyes on the opponent, arches his back, and jumps at his opponent, attempting to bite. The opponent, on the other hand, moves his tail—bent into an S-shape—excitedly from one side to the other and then flees. The mating season for this southern European species is in spring. The male appears to stimulate the female by touching her body with his snout (similar to *Hemidactylus flaviviridis*).

About 15 to 25 days after copulation the female deposits 2 eggs (about 1.2 x 1.0 cm, weighing 0.6 grams) in cracks in walls, among cork bark, or sometimes even in sand. Initially the shells are still parchment-like, but they harden to a chalk-like consistency within the next few hours. Within one reproductive cycle several eggs can be laid in 2 to 3 week intervals. The eggs should be removed, their upper sides properly marked, and then with the substrate be placed into a Type II brood container. Direct contact between eggs and water must be avoided. The incubation period lasts about 50 days at a temperature of 28-31° C. The young are about 35 mm long at birth and should be transferred into a dry terrarium. The initial diet consists of small insects. Sexual maturity is reached after 1 year.

The following geckos can be kept and bred under identical conditions:

Alsophylax pipiens
Maximum size: 12 cm.
Distribution: Afghanistan to China.
Habitat: Buildings, rocky areas.
Clutch size: 2.
Incubation period and hatchling size: 44 to 53 days at 20 to 28° C; 3.5 cm.

Cyrtodactylus caspicus
Maximum size: 16.3 cm.

294

Distribution: Northwestern shores of Caspian Sea, eastern Trans-Caucasus, Central Asia.
Habitat: Buildings and walls in arid regions.
Clutch size: 2.
Incubation period and hatchling size: 81 to 168 days at 19 to 26° C; 4.5 cm.

Cyrtodactylus kotchyi
Maximum size: 13.5 cm.
Distribution: Southeastern Italy, Balkans, Asia Minor, Syria, Lebanon, Israel, Cyprus.
Habitat: Buildings, walls, trees in semi-arid region.
Clutch size: 2.
Incubation period and hatchling size: 78-83 days at 23-28° C; 3 cm.
Remarks: Sexual maturity in 2 years.

Phyllodactylus europaeus
Maximum size: 8 cm.
Distribution: Islands in western Mediterranean Sea and those off Tunisia and Italy.
Habitat: Buildings, rocky arid regions.
Clutch size: 2.
Incubation period and hatchling size: 85-86 days at 23° C; 3 cm.
Remarks: Sexual maturity in 3 years.

Iguana iguana—Green Iguana
(Family Iguanidae)
Description: Size to 180 cm. Dorsum light to dark green, frequently with grayish green to brown crossbands. Ventral area yellowish green. Strong legs with clawed toes. Spiny crest from nape to first third of tail. Large tuberculate scales in the posterior section of the lower jaw (throat region), one scale especially large. Males with stronger spinous crest and better developed thigh pores.
Distribution and habitat: Tropical Central and South America, Antilles, Trinidad. On trees and bushes close to water.
Care: Food: Insects, newly born mice, beef heart, fruits, vegetables. Should be kept in a semi-moist terrarium at air temperatures from 25 to 38° C during the day and 18 to 22° C at night. Use localized heating with a radiator of some type plus

Day geckos. **Top:** *Phelsuma madagascariensis.* **Bottom Left:** *Phelsuma mada-gascariensis.* Photo by H. Zimmermann. **Bottom Right:** *Phelsuma laticauda.* Photo by B. Kahl.

Top: *Ptyodactylus hasselquistii*, the fan-fingered gecko. Photo by E. Zimmermann. **Bottom:** *Physignatus lesueurii*, the eastern water dragon. Photo by B. Kahl.

UV radiation. The cage should be sprayed several times daily. Sand or a peat moss and sand mixture is good as a substrate. Needs several strong climbing branches. Provide a large water section.

Behavior and breeding: Males of this species are commonly territorial, particularly so during the breeding season. At that time ritualistic fights among males are a common occurrence. During such fights both males raise their bodies on all four legs and spread their dark-colored gular fold with the aid of the hyoid bone. With a curved back they turn their dorsoventrally compressed body toward the opponent. If the opponent does not respond with a submissive position (body firmly pressed against the ground) but instead also shows the threat display, both animals will begin to circle each other. Then, facing each other head on, they thrash their heads laterally together until one of them gets tired and turns away. Usually the defeated animal is not followed; in fact, the dominant male may even tolerate the subordinate one in the same terrarium. However, mating is reserved for the dominant male.

Females are usually less aggressive, and normally they only defend their sleeping place against other females. However, in a confined situation such as a small island there can also be fights among females. For instances, along a narrow shoreline there are only a limited number of egg-laying sites available. Females dig nest burrows 1 to 2 m long and about 60 cm deep into the damp sand, where eventually 20 to 30 eggs are deposited. According to natural habitat studies these burrows may have incubation temperatures of 20 to 32° C. By marking newly hatched iguanas on a small island in the Panama Canal it has been observed that these animals are capable of coordinating their activities. All young from a particular clutch hatch at nearly the same time, burrow up to the surface, and then migrate jointly by swimming across the lake to the mainland close by. There they tend to live primarily among low brush vegetation. Several animals will go hunting together and may even sleep on top of each other among the branches.

Green iguanas have been repeatedly bred in captivity in large terraria in recent years. The male bites firmly into the neck or head of the female and grasps around the female's tail with one hind leg during copulation, which can last from

1 to 20 minutes. While in that position the male's head sways rhythmically from one side to the other.

Gestation varies from 49 days to 90 days. About 7 to 40 days prior to egg-laying the female's food intake stops and an increasing amount of water is taken up. At a site with damp sand the female digs a hole and lays 20 to 30 eggs within a period of about 5 hours. The eggs are laid in pairs at intervals that become increasingly longer. The eggs are approximately 34 to 35.5 mm long, with a diameter of 24 to 24.6 mm and a weight of 11.4 grams. After all eggs have been carefully marked at their upper end they are transferred to a Type I brood container. Damp sand and peat moss are suitable substrates for incubation. The incubation period lasts on the average 113 days at a temperature of 28 to 30° C. At birth the emerald-green young are 19 to 27 cm long. They must be transferred to a semi-moist terrarium with plenty of climbing facilities and decorated as for adults. The first food is taken after 10 days. The diet should consist of clover, dandelion and lettuce leaves, finely ground commercial turtle food enriched with calcium supplement, and ground carrots. UV radiation should be given three times a week for 5 minutes each with Osram-Ultravitalux. Sexual maturity is reached in 2 to 3 years.

Green iguanas, even those in captivity, are keen observers and can adjust to different situations. Therefore, their visual learning capacity was tested experimentally. This led to the discovery that this species is capable of distinguishing between 2 approximately complimentary colors such as red-green, blue-yellow and black-white, as well as between test pairs, such as narrow versus wide stripes, green dots against brown wavy lines, and similar patterns. Such acquired knowledge can be retained in memory for up to 5 months. In comparison to lizards with smaller brain and body size, the learning and retention capabilities are thus substantially larger.

Ctenosaurus similis, the **Black Iguana** *(1.1 m; Central America; grassland, trees), can also be kept and bred under identical conditions in large terraria. The incubation period is about 98 days at 26 to 29° C. The young are 14 cm long at birth.*

Top: *Teratoscincus scincus*, the wonder gecko. Photo by B. Kahl. **Bottom:** *Sceloporus jarrovi*, Yarrow's spiny lizard. Photo by B. Kahl.

Top: *Tiliqua scincoides*, the eastern blue-tongued skink. Photo by E. Zimmermann. **Bottom:** *Trachydosaurus rugosus*, the shingleback or pinecone skink. Photo by E. Zimmermann.

Lacerta lepida—Pearl Lizard, Eyed Lizard
(Family Lacertidae)

Description: Size in excess of 60 cm. Dorsum green, gray, or brownish with black dots. Flanks usually with conspicuous blue spots. Ventral area yellow to greenish. Male with broader head and well developed femoral pores.

Distribution and habitat: Iberian Peninsula, southern France, northwestern Italy, and northwestern Africa. Lowlands to mountain regions in sun-exposed, dry regions with brush vegetation.

Care: Food: Worms, insects, snails, small mammals (rodents), small birds, strips of beef heart. Should be kept in a dry terrarium or in an outdoor enclosure during the summer months. Air temperatures 25 to 40° C during the day and 18 to 24° C at night. Localized heating with a radiator and regular UV radiation are necessary, as is misting daily with water. Use sand as a substrate. Decoration should include rock structures and climbing branches, as well as plants with hard leaves, succulents, and similar species (plants can be omitted). This species should be over-wintered at about 10° C.

Behavior and breeding: If threatened, eyed lizards display a characteristic defensive behavior. They open their mouth wide, bend their body into an S-shape, and emit a hissing sound created while exhaling when the tongue is pulled back and the opening to the trachea becomes restricted. When directly approached they will jump at any aggressor, bite, and then immediately run away. Since the males have particularly strong jaws that can cause bleeding wounds, caution has to be shown when this species is handled.

If a new male is added to a terrarium that already has a male, the resident territory holder will attempt to intimidate the newcomer by taking up a threat posture. He inflates his gular sac and positions his body obliquely to the opponent so the many blue ocelli are toward the opponent. This usually causes the intruder to either flee or indicate his submissiveness by nodding his head and twitching his tail. Real fights rarely take place.

During the breeding season a female willing to mate will maintain visual contact with a male. If the male approaches the female remains motionless. The male then moves closer, licks the flanks and cloacal region of the female, and places his head on or next to the female's hind legs. He flicks his tongue again, moves back slightly, and finally grasps one of the female's hind legs, her flank, or the base of the tail in his jaws. Once the mating bite has been correctly placed on her flank, copulation occurs during the next 8 to 15 minutes. The male lies parallel and directly next to the female, his hind leg grasping her tail base and with his cloacal region pressed against hers. Following mating the female refuses all other males with defensive and threatening displays, such as raising the neck region, expanding the gular fold, head bobbing, biting threats, and tail twitching.

The eggs (6 to 23) are deposited in a damp substrate area. The clutch should be treated the same way as described for *Lacerta viridis*. The incubation period lasts from 70 to 85 days at an incubation temperature of 25 to 28° C inside the brood container. Newly hatched specimens have a characteristic juvenile pattern: they are grayish green with numerous yellowish (often with a dark margin) ocelli distributed over the dorsal areas, and they have a relatively large head. The juveniles are raised in a dry terrarium where they should be given a varied, vitamin-enriched diet of insects and similar foods as well as calcium supplements such as crushed cuttlebone. They need regular UV radiation. They become sexually mature after about 3 years.

The following species can be kept and bred under the same conditions:

Lacerta jayakari
Maximum size: 44 cm.
Distribution: United Arab Emirates.
Habitat: Dry areas, dry river beds.
Clutch size: 9.
Incubation period and hatchling size: 116-119 days at 27-30° C; 17.1 cm.

Lacerta pater
Maximum size: 60 cm.
Distribution: Northwestern Africa.
Habitat: Dry regions.
Clutch size: 20.
Incubation period and hatchling size: 88-89 days at 29° C; 12 cm.

Top: *Varanus storri*, a pygmy monitor. Photo by H. Frauca. **Bottom:** *Varanus timorensis*, the spotted tree monitor. Photo by H.-G. Horn.

Top: *Chondropython viridis*, the green tree python, young. Photo by B. Kahl. **Bottom:** *Boa constrictor*, the common boa or boa constrictor. Photo by B. Kahl.

Remarks: Sexual maturity after 1 year.

Lacerta simonyi
Maximum size: 60 cm.
Distribution: Grand Canary Island.
Habitat: Dry regions.
Clutch size: 11.
Incubation period and hatchling size: 65-72 days at 28-32° C; 16.5 cm.

Lacerta strigata
Maximum size: 35 cm.
Distribution: Asia Minor, Trans-Caucasus, Arabia.
Habitat: Dry grasslands.
Clutch size: 9.
Incubation period and hatchling size: 44-66 days at 28-32° C; 7 cm.

Lacerta viridis—Green Lizard
(Family Lacertidae)
Description: Size to 40 cm. Dorsum green to brown with numerous black dots or four whitish dorsal stripes. Belly yellowish. Males with enlarged femoral pores and deep blue throat during mating season. Hybridization with *Lacerta agilis* and *L. trilineata* possible.
Distribution and habitat: Central and southern Europe to Asia Minor. Lowlands to mountain regions, in dry, sun-exposed to semi-moist areas with abundant brush vegetation. Frequently close to water.
Care: Food: Insects, spiders, worms, fruit, honey. Should be kept in a semi-moist terrarium or an outdoor enclosure during the summer months at air temperatures from 25 to 35° C during the day and 18 to 22° C at night; localized heating with a radiator plus regular UV radiation is best. A mixture of leafy soil and peat moss will do as a substrate (half damp, half dry). Decorations can include climbing branches and rock structures. Use plants with hard leaves, such as agaves or similar species. Give a small water bowl. This lizard should be over-wintered at about 10° C.
Behavior and breeding: Green lizards will establish a hierarchial system—according to the sexes—in a large, well-laid out, and properly decorated terrarium. This becomes partic-

306

ularly apparent during the breeding season. The position of an individual within the hierarchy is dependent upon body size, color intensity, and aggressiveness.

During the breeding season the alpha (dominant) male is especially brightly colored with a bright green back and dark blue throat. It runs frequently through the terrarium with its head held high to indicate its dominance to all other members of the community. Subordinate males will then signal their submissiveness with "leg kicking," "head nodding," and moving the forelegs alternately up and down while the abdomen is firmly pressed against the ground, remaining motionless. This behavior pacifies the dominant male, which then turns away.

However, if the lead male encounters another male that neither turns away nor displays the submissive behavior, this will lead to threat displays. To do that the alpha male will run toward the other male in spurts with his head and upper torso slightly raised. It will expand the dark blue gular fold and move its tail laterally a few times. If the other male then also begins to display open threats, a fight will ensue during which both opponents encircle each other, biting into the other one's tail.

The strongest male is usually the only one that reproduces. He will approach a female, trying to impress her with his body displays. He then circles the female and finally stops her by biting into her tail. Although she initially offers some modest resistance, this fades quickly. She stops so that the male can now change position. He moves slightly further up, applies his mating bite laterally into the flank of the female, and grasps the female's tail base with one hind leg. This is followed by copulation lasting about 5 minutes, during which the male ignores other males, but the female bites other males that are trying to court her during copulation.

About six weeks later the female begins to search for a damp site in the terrarium, where she digs a hole. There she deposits 6 to 21 eggs (15-18 mm x 11-18 mm). The eggs should be removed, their upper side properly marked, and then be transferred to a plastic container filled with a damp mixture of sand and peat moss, which in turn is then placed inside a Type I brood container. The young hatch in about 57 days at an air temperature from 26 to 35° C. At birth they are 6 to 9 cm long and weigh 3 to 4 grams. Compared with

Top: *Coluber constrictor foxi*, a blue racer. Photo by J. K. Langhammer. **Bottom:** *Coluber viridiflavus*, the Eurasian green and yellow racer. Photo courtesy Dr. D. Terver, Nancy Aquarium, France.

Elaphe guttata, the corn snake. Photo by B. Kahl.

the rest of the body, the head is initially relatively large. The juvenile coloration is brownish, frequently with two to four lighter dorsal bands. The young should be raised in a semi-moist terrarium and be given a vitamin-rich diversified diet of small insects. Regular UV radiation is essential. Sexual maturity is reached in about 3 years.

The following species can be kept and bred under the same conditions:

Lacerta schreiberi
Maximum size: 35 cm.
Distribution: Iberian Peninsula.
Habitat: Vegetation along water.
Clutch size: 21.
Incubation period and hatchling size: 42-47 days at 28-29° C; 7 cm.

Lacerta agilis
Maximum size: 20 cm.
Distribution: Northern and central Europe to central Asia.
Habitat: Forests, yards, along agricultural land.
Clutch size: 15.
Incubation period and hatchling size: 35 to 36 days at 31° C; 3.5 cm.
Remarks: Over-winter at 5° C.

Lacerta praticola
Maximum size: 15 cm.
Distribution: Eastern Yugoslavia to Caucasus.
Habitat: Forests, paddocks.
Clutch size: 6.
Incubation period and hatchling size: 60 days at 23-27° C; 3 cm.

Lygodactylus picturus—Dwarf Gecko
(Family Gekkonidae)
Description: Size to 7.2 cm. Dorsum bluish gray with yellow-brown marbled pattern on head. Abdomen white to yellow with yellow-orange median band. Two rows of adhesive lamellae on tip of tail. Male with black throat and enlarged pre-anal pores.

Distribution and habitat: Eastern Africa. Semi-moist areas, on palm trees, acacias, or on fences and building walls close to human habitation.

Care: Food: Insects, spiders, sweet fruit. Should be kept in a semi-moist terrarium with air temperatures from 25 to 32° C during the day and 18 to 22° C at night. Supply localized heating with some type of radiator plus regular UV lighting. Daily water spraying a must. Use a mixture of sand and peat moss or fresh moss as a substrate and cork bark for the back wall. Decorations can include climbing branches, *Pandanus*, *Cissus*, or similar plants. A small water bowl should be available.

Behavior and breeding: This gecko is active mainly during the day. Males will establish territories in a terrarium just as they do in the wild. The size of these territories is dependent upon the set-up of the terrarium and the size of the individual males involved. Violations of territory boundaries will invariably lead to aggressive actions. First come display threats in order to intimidate and drive off any trespasser. If this does not chase the intruder away or cause him to display a submissive posture, the resident male will jump at him, trying to bite. This is followed by a chase throughout the terrarium that only stops when one of the opponents presses his body firmly against a branch and assumes a grayish brown submissive coloration or if one animal withdraws out of sight of his pursuer. Subordinate animals that are constantly chased and so have to remain in hiding and are essentially unable to feed properly should be removed into separate quarters in order to avoid mortalities.

In captivity the breeding period does not appear to be seasonal, since males will court throughout the year. Courtship commences with the male approaching a female to within 20 cm, signalling threat displays. If the female is not willing to mate she launches a simulated attack and also threatens. However, if she does not move the male approaches her from behind and flicks his tongue at the tip of her tail or simply bites into one of her hind legs in the thigh area, pushes his body over hers, and attempts to secure a firm mating bite at her neck. The female's head begins to tremble and she raises her tail (now held in an S-shape) slightly so that the male can press his cloacal region against hers. During the subsequent copulation, which can last up to 32 minutes, the male's color-

Eryx johnii, the Indian sand boa. Photo by B. Kahl.

Top: *Lampropeltis getulus getulus*, the common kingsnake or chainsnake. Photo by Ken Lucas at Steinhart Aquarium. **Bottom:** *Lampropeltis zonata*, the California mountain kingsnake. Photo by B. Baur.

ation becomes more subdued. The eggs are usually deposited in cracks and crevices, preferably among pieces of cork bark used for decoration. Two white eggs (about 0.5 x 0.35 cm in size, usually adhering to each other) are laid at a time. They will harden following exposure to air. Shortly after having laid the eggs the female will often copulate again, and 17 to 21 days later another clutch is produced.

The eggs, including the substrate to which they are attached, should be placed in a Type II brood container. The incubation period lasts on the average 78 days at a temperature of 25 to 32° C. At hatching the young geckos are quite small (24 to 28 mm) and display a grayish brown juvenile coloration with white spots. For better control and development they should be transferred to a clear plastic refrigerator container with a gauze lid, a damp foam rubber bottom, some pieces of cork, and a *Scindapsus* runner. The initial diet should consist of newly hatched crickets, fruitflies, small wax moth larvae, and similar items. They should also be given daily UV radiation with UV-A lamps. The coloration begins to change after 2 months, and by 6 months the young are indistinguishable from their parents. At that stage they are nearly sexually mature.

Mabuya quinquetaeniata—African Blue-tailed Skink (Family Scincidae)

Description: Size to 30 cm. In females and juveniles dorsum black with cream longitudinal stripes, tail blue, abdomen white. Dorsum in males usually brown with two wide reddish brown bands and a narrow light-brown median dorsal stripe; dorsal surface of tail brown to red, ventrally yellow; abdomen yellow; head and neck with white to light blue spots arranged in longitudinal row.

Distribution and habitat: From Mali to Egypt, south to South Africa. Savannah, steppes, gardens. On bushes, in sandy areas with sparse plant growth, on rocks, walls, and roofs.

Care: Food: Insects, worms. Should be kept in a dry terrarium at air temperatures from 25 to 45° C during the day and 18 to 22° C at night. Supply localized heating with a radiator and floor heating as well as regular UV radiation. Coarse river sand (one half damp, the other dry) makes a good substrate. Decorations can include rocks, branches, or twigs. Succulents and desert grasses may be planted. Give a small water dish.

Behavior and breeding: This skink is active mainly during the day. It is a ground-dwelling species that exhibits a more or less defined social structure, a pair together with their young occupying a particular territory. This interestingly colored skink has been bred repeatedly in captivity. In a terrarium the female buries up to 11 eggs (15 mm long) at a damp location. Once the eggs have been deposited, the hole is meticulously covered over again with soil.

The upper side of the eggs should be marked and the eggs then transferred into a clear plastic container (Type I brood container) filled with slightly damp sand. At an average air temperature of 29° C the 6 to 8 cm young will hatch after an incubation period of about 36 days. At that time they already possess the attractive dark blue tail. The young are transferred into a dry terrarium and are given a vitamin-enriched diet of small insects (they begin feeding a few hours after hatching) as well as some crushed calcium (cuttlebone). Regular UV radiation is also essential.

Care and breeding of the following skinks conform largely to the above species:

Cophoscincopus durus
Maximum size: 12 cm.
Distribution: West Africa.
Habitat: Along water, under leaves, rocks or wood.
Clutch size: 2.
Incubation period and hatchling size: 49 days at 20-22° C; 3.8 cm.
Remarks: Start diet for young with springtails.

Mabuya aurata
Maximum size: 28 cm.
Distribution: Eastern Africa to Afghanistan and Pakistan.
Habitat: Open grasslands and bush vegetation.
Litter size: 8; live-bearing.

Mabuya capensis
Maximum size: 20 cm.
Distribution: South Africa.
Habitat: Savannahs, deserts with sparse vegetation.

Top: *Lampropeltis triangulum*, a milk snake or scarlet kingsnake. Photo by B. Kahl. **Bottom:** *Natrix natrix*, the ringed snake or Eurasian grass snake. Photo courtesy Dr. D. Terver, Nancy Aquarium, France.

Right: *Natrix maura*, the viperine water snake. Photo by E. Zimmermann. **Bottom:** *Python molurus*, the Indian python. Photo by Dr. S. A. Minton.

Litter size: 15; live-bearing.
Juvenile size: 7.5 cm.

Mabuya vittata
Maximum size: 18 cm.
Distribution: Northeastern Africa, Asia Minor, Syria, Lebanon.
Habitat: Grasslands with brush vegetation.
Litter size: 4; live-bearing.
Juvenile size: 7 cm.
Remarks: Very aggressive, should be kept in pairs when adult; rear in small groups.

Riopa fernandi
Maximum size: 34 cm.
Distribution: Western Africa.
Habitat: Rain forest.
Clutch size: 9.
Incubation period: 56 days.
Remarks: Should be kept in a moist terrarium, the substrate, at least 10 cm deep, being mixture of sand and peat moss planted with *Scindapsus, Hoya,* or similar species.

Oedura monilis—Ocellated Velvet Gecko
(Family Gekkonidae)
Description: Size to 16 cm. Dorsum dark gray with light gray and yellow spotted or marbled pattern. Ventral area white to whitish yellow. Male with enlarged cloacal region and pre-anal pores.
Distribution and habitat: Eastern Australia. Semi-moist regions, beneath the bark of dead trees.
Care: Food: Insects, spiders. Should be kept in a semi-moist terrarium at air temperatures of 22 to 25° C during the day and 18 to 22° C at night. Spray with water in the evening. The back wall can be made up of cork bark, and decorations can include climbing branches and plants such as *Hoya, Ficus, Acacia,* or similar species. A small water dish should be available.
Behavior and breeding: Velvet geckos are nocturnal animals. The adhesive lamellae on the fingers and toes equip them very well for a life in the trees. However, they will also follow their prey down to the ground, where they may travel over

318

long distances. In their natural habitat one male lives together with a female and their progeny in a distinct territory that is defended against any male intruder. The minimal distance between males in adjacent territories must be about 60 cm.

Sexually mature males maintain clearly defined, well-delineated territories even in large terraria. If one male trespasses into the area of another it will be closely watched by the resident male. If the trespasser continues to approach he will be warned with threat displays, such as raising the body on all four legs and arching the back. The excitement and tension are best reflected in the sinusoidal tail movements of the resident male. If this does not intimidate the intruder, that is, if he does not press his body firmly against the ground (submissive posture) or flee, but he instead continues to approach, the resident male attacks. He lunges toward the intruder and tries to grab him on the head, neck, or tail. If serious injuries to either opponent are to be avoided, the males should be separated immediately. Therefore, it is advisable to keep this species only in pairs or only together with their sexually immature progeny.

The *Oedura* female produces clutches of 2 parchment-like white eggs at intervals of 1 to 3 months. They are laid on damp spots in the terrarium, preferably in damp peat moss or sand in flowerpots. The eggs are 21 mm long with a diameter of 4 mm and a weight of 1.1 grams. We have left some of the eggs in the terrarium, but it is advisable that they be removed after the upper side has been clearly marked. The eggs are then placed in sterilized, damp sand inside a clear plastic container that in turn is kept inside a Type I brood container. As do all soft-shelled reptilian eggs, they tend to pick up water during the course of their development and thus increase in size and weight with time.

The incubation period lasts about 77 days at an air temperature of 25 to 32° C. At hatching the young are 75 mm long and are still lead-gray. The first molting occurs just half an hour after hatching and takes about 1 hour. Only then can we see the attractive juvenile coloration, which is velvet-black with many yellow dorsal bands. These young geckos will feed from the third day onward, searching for small crickets and fruitflies. With a varied vitamin-rich diet the young will grow quickly, reaching sexual maturity at an age of 1 year.

Top: *Thamnophis sirtalus tetrataenia*, a strikingly colored garter snake. Photo by Ken Lucas at Steinhart Aquarium. **Bottom:** *Thamnophis elegans*, the western terrestrial garter snake. Photo by B. Kahl.

> *Oedura marmorata,* **Marbled Velvet Gecko** *(15 cm; Australia; trees and rock crevices), can be bred under the same conditions. The female produces 2 parchment-shelled eggs. The incubation period lasts 51 to 87 days at a temperature of 25-32° C. The young are 7 cm long at birth.*

Ophisaurus apodus—Scheltopusik
(Family Anguidae)

Description: Size to 125 cm. Dorsum yellow to dark brown, abdomen lighter. Rough scales supported by bony plates; deep, non-bony lateral fold; tiny hind legs. Snake-like appearance.

Distribution and habitat: Southeastern Europe, Caucasus, Asia Minor to central Asian Soviet Republics. Sun-exposed areas in fields and paddocks, among hedges and bushes as well as under rock piles.

Care: Food: Insects, snails, mice, nestling birds, eggs, pieces of meat. Should be kept in a dry terrarium or in an outdoor enclosure during the summer months at air temperatures of 25 to 30° C during the day and 18 to 22° C at night. Provide localized heating with a radiator, regular UV radiation, and a daily water spray. Half of the substrate area should be sand, the other half leafy top soil. Decorations should include large, flat stones and roots as hiding places. Provide a sturdy water pan. It is recommended that this species be over-wintered in a mixture of leafy soil and moss at about 10° C.

Behavior and habitat: This species is active throughout the day, hunting for its prey exclusively on the ground. Once it has discovered prey, it approaches slowly, elevating the head slightly, flicking its tongue, and lunging forward with its mouth wide open. The prey is grasped, killed by pressing it against the ground, and finally swallowed.

If one of these almost legless lizards is caught it will attempt to scare off the potential enemy by discharging the foul-smelling contents of its digestive tract and at the same time trying to force its release by twisting the body around its axis.

Breeding this species requires a large terrarium of at least 200 x 200 x 100 cm. About 2 months after mating, which occurs during late spring, the female deposits 6 to 10 white

321

soft-shelled eggs (4 x 2 cm) at a damp site. After they have been properly marked they can be transferred in a Type I brood container.

The young will hatch after about 6 weeks. At birth they are 10 cm long. The characteristic juvenile pattern is a light gray back with dark brown crossbands. They should be transferred to a dry terrarium, where they are fed small insects, snails, and pieces of raw meat.

Gerrhonotus multicarinatus, **Southern Alligator Lizard** *(43 cm; western USA, semi-wet areas among grassland and in open forests),can be bred in very large terraria under identical conditions. The female produces 2 to 3 clutches of eggs per year for a maximum total of 41 eggs (size 17 x 22 mm). The female provides brood care. Better hatching results are obtained with artificial incubation (brood container Type I) in damp sand at a temperature of 27 to 29° C. The incubation period is 40 to 44 days. The young are 9.7 cm to 10.2 cm long at hatching.*

Phelsuma lineata—**Striped Day Gecko**
(Family Gekkonidae)
Description: Size to 15 cm. Dorsum green with red-brown spots distributed more or less over the area. A mostly brown-black lateral band from nose to base of tail. Ventral side white. Males with pre-anal and femoral pores as well as well-developed postanal glands.
Distribution and habitat: Madagascar. Along the edges of evergreen forests, on banana plants, brush, bushes, often close to human habitation.
Care: Food: Insects, spiders, sweet fruit, jams, honey, flower nectar. Should be kept in a moist terrarium at air temperatures from 25 to 35° C during the day and 15 to 20° C at night. Localized heating with some type of radiator as well as regular UV radiation is desirable. Spray in the morning and evening. Use a mixture of leafy soil and peat moss as a substrate. Decorations can include climbing branches and plants that offer vertical and oblique crawling areas. A small water bowl should be available.
Behavior and breeding: As in all *Phelsuma* species, the striped

322

day gecko is active primarily during the day. Since the males are highly territorial in captivity, it appears safe to assume that they behave similarly in the wild. Aggression among these animals, particularly the males, is quite common in the confines of a terrarium. If one male sees another one there is the inevitable initial threat display. The resident male displays its conspicuously colored dorsal area obliquely to the intruder. At the same time the resident male flicks its tongue and whips its tail excitedly. If the intruder then retreats or takes on a darker coloration, the resident male turns away. However, if the intruder also presents the lateral threat display, a fight ensues. Both males circle each other and attempt to bite. In view of the restricted area available in a terrarium, these intra-specific fights can lead to substantial injuries, including the loss of toes or the tip of the tail. Defeated animals often continue to be intimidated and threatened by the more dominant ones, so they may not be able to feed properly and thus lose condition. These suppressed males should be transferred to a separate terrarium. In fact, smaller terraria (50 x 30 x 60 cm or so) should really only hold one pair.

During the breeding season the male approaches the female with the characteristic lateral threat display. He climbs several times over the female, his head trembling. He finally applies the mating bite to the female's neck, and copulation follows.

The female ceases to feed about 2 weeks prior to laying the eggs and begins to search for a suitable site to deposit the eggs. Eventually two eggs are expelled in succession accompanied by lateral contractions of the body. The hind legs are folded together to catch the eggs, turning them several times and finally pressing them together and pushing them firmly against the substrate. About 2 hours later the female resumes feeding again. Up to 6 clutches of 2 eggs each can be laid by one female in the course of a year.

The eggs are 10-12 mm in size and will quickly harden after having been attached to the substrate. They should be removed together with a portion of the substrate or left in the terrarium but protected with a plastic or wire gauze cover. After the eggs have been properly marked on the upper side they should be transferred to a Type II brood container. The incubation period lasts about 45 days at a temperature of 26 to 28° C. At the time of hatching the young are about 40 to

50 mm long. They still retain the embryonic skin, but this skin dries quickly, turns silver-gray, and shortly thereafter the first molt begins. To facilitate shedding, the young gecko rubs himself along the substrate or some other object. The last remnants of the old skin are pulled off with his mouth. Subsequently the entire shed skin is eaten. Now the juvenile coloration is fully visible, the green dorsal region having a multitude of red to black dots. From that point on the requirements are the same as for the adults. The diet should be kept varied, consisting of insects and a mixture of sweet fruit, honey, or marmalade (it is easy to mix in multivitamin and mineral preparations). Regular UV radiation is also essential for rearing the young gecko.

Care and breeding of the following gecko species are largely similar to that for the species mentioned above:

Phelsuma abbotti
Maximum size: 15 cm.
Distribution: Seychelles, northern Madagascar.
Habitat: Forests.
Clutch size: 2.
Incubation period and hatchling size: 85-90 days at 28° C; 4.9 cm.
Remarks: Deposits eggs in cracks and crevices.

Phelsuma astriata
Distribution: Seychelles.
Habitat: Forests.
Clutch size: 2.
Incubation period and hatchling size: 35-40 days at 28° C; 3.7 cm.

Phelsuma cepediana
Maximum size: 14 cm.
Distribution: Mauritius, Reunion.
Habitat: Damp forests, palm trees, buildings.
Clutch size: 2.
Incubation period and hatchling size: 35-45 days at 27-29° C; 4.5 cm.
Remarks: Eggs deposited on leaves and in cracks or crevices in tree bark.

Phelsuma dubia

Maximum size: 15 cm.
Distribution: Madagascar, Comoro Islands, Zanzibar, Tanzania.
Habitat: Forests, palm trees.
Clutch size: 2.
Incubation period and hatchling size: 50 days at 28° C; 4.7 cm.
Remarks: Eggs deposited in tree bark, wood.

Phelsuma guimbeau

Maximum size: 17 cm.
Distribution: Mauritius.
Habitat: Damp forests.
Clutch size: 2.
Incubation period and hatchling size: 75 days at 27 to 29° C; 5 cm.

Phelsuma laticauda

Maximum size: 12 cm.
Distribution: Eastern Madagascar, Comoro Islands, Nossi Be.
Habitat: Damp forests, plantations.
Clutch size: 2.
Incubation period and hatchling size: 37-52 days at 28° C; 4 cm.
Remarks: Eggs deposited on *Sanseveria* leaves.

Phelsuma madagascariensis

Maximum size: 26 cm.
Distribution: Madagascar and surrounding islands.
Habitat: Damp forests, palm trees, buildings.
Clutch size: 2.
Incubation period and hatchling size: 60-65 days at 30° C; 6 cm.
Remarks: Eggs deposited in among cork bark; sexual maturity in 1 year.

Phelsuma ornata

Maximum size: 12 cm.
Distribution: Mauritius.
Habitat: Semi-damp areas; among bamboo.
Clutch size: 2.
Incubation period and hatchling size: 40 days at 28° C; 4 cm.

Remarks: Eggs deposited on leaves and tree bark.

Phelsuma standingi
Maximum size: 25 cm.
Distribution: Southwestern Madagascar.
Habitat: Fringing forests, bushland.
Clutch size: 2.
Incubation period and hatchling size: 64-73 days at 28-30° C; 6.6 cm.

Gonatodes albogularis
Maximum size: 10 cm.
Distribution: Lesser Antilles, Florida, Central America to northern South America.
Habitat: Buildings, walls, damp forests.
Clutch size: 1.
Incubation period and hatchling size: 58-72 days at 28° C; 3 cm.
Remarks: Deposits eggs in pits; sexually mature in 6 to 7 months.

Physignatus lesueurii—Eastern Water Dragon
(Family Agamidae)
Description: Size to 100 cm. Dorsum brownish gray with black to horn-colored spots. Dark band from posterior eye margin to base of foreleg. Abdomen brick red to light brown. Serrated scaly crest with particularly large spines along nape.
Distribution: Eastern Australia and New Guinea. Tree-covered river banks, also extending into brackish water regions.
Care: Food: Insects, worms, crabs, fish, newly hatched chicks, mice, fruit, small pieces of meat. Should be kept in a semi-moist terrarium at an air temperature ranging from 25 to 28° C during the day and 20 to 22° C at night. Must be sprayed during the morning hours. Localized heating with radiator plus regular UV radiation must be supplied. Half the bottom should be sand, the other half water with level to about 25 cm depth. Decorations can include rough climbing branches as well as *Ficus, Hoya,* and similar species (out of reach of the animals), but can also be omitted.
Behavior and breeding: According to observations in the wild, this species is easier heard than seen! If a flight distance of 6

m is exceeded, these animals will let themselves fall off a branch and into the water below, sometimes from as high as 9 m! The eastern water dragon is a good swimmer; with its limbs held close against the body the tail is used as a rudder for propulsion. The snout may be just barely above the surface or the animal may swim completely submerged until it is safe to surface again, usually among bushes along the river bank. There the animal remains until the danger has passed. Water dragons escaping over land utilize a technique already described for the South American basilisks: they raise the body on the hind legs and race in this bipedal fashion. However, as soon as a vertical surface has been spotted they will climb up.

If a water dragon is caught it will kick its hind legs backward and use the tail in a whip-like manner. Actual bites do occur, although this happens rarely; bites should be avoided since such bites and scratch wounds from the sharp claws have a tendency to heal slowly, a point that should be taken into consideration when *P. lesueurii* is handled.

Water dragons have a well-defined territory that each animal crisscrosses several times in the course of a day. In the wild this includes trees and a section along the river bank. If another water dragon trespasses the territory boundaries, the resident male nods several times with his head, inflates his gular fold, and moves his arms alternately in circles and up and down. With this territory indicator the intruder is informed of the territorial claim of the resident.

Little is known about the reproductive behavior of this species, although captive breeding (as well as that of the related Asian water dragon, *Physignatus cocincinus*) has been successful in large terraria. The female buries 8-18 oval, parchment-like eggs (frequently several times a year) in an area with loose, damp substrate. The eggs should be carefully removed, their upper side properly marked, and then be placed in a plastic container with damp sand to be incubated in a Type I brood container. The incubation period lasts about 86 days at a temperature of 30° C. At the time of hatching the young are approximately 12.7 cm long. They are moved to a semi-moist terrarium and fed insects. A vitamin-rich, varied diet including occasional mineral supplements is essential, together with regular UV radiation.

> *Care and breeding of* **Physignatus cocincinus** *(100 cm; India to eastern Asia, lowlands and highlands close to water) correspond to that for the previous species. The female produces up to 16 eggs that hatch after 67 to 101 days at an incubation temperature of 29-30° C.*

Podarcis sicula—Ruins Lizard
(Family Lacertidae)

Description: Size to 25 cm. Dorsum green, yellowish olive, or light brown with a black reticulated pattern (which can also be absent); dorsal markings highly variable. Ventral region whitish to green or blue, usually without spots. Male with clearly developed femoral pores.

Distribution and habitat: Italy and Dalmatia. Isolated colonies on Minorca and in Istanbul; a few introduced colonies in the USA. Lowlands to plateaus of 1200 m elevation; along walls, cliff faces, ruins, and in bushes.

Care: Food: Insects, worms, sometimes also plants, vegetables. Should be kept in a dry terrarium or in an outdoor enclosure during the summer months at air temperatures of 25 to 40° C during the day and 15 to 22° C at night. Must be sprayed daily. Provide localized heating with a radiator and regular UV radiation. Use sand or a mixture of sand and peat moss as a bottom substrate. Decorations can include climbing branches and rocks as well as *Sedum, Sempervivum,* and similar plants. Use a small water dish. Over-wintering at 8 to 10° C is recommended.

Behavior and breeding: The males of this species display a well-defined territoriality. If one male trespasses into the territory of another, there is usually an agonistic interaction. The resident attempts to intimidate the intruder by a threat display—i.e., lowered head, inflated throat pouch, and laterally flattened body. If the intruder does not signal his submissiveness by "leg kicking" ("treteln"), alternately raising and lowering the forelegs and trembling with the tail, or if he does not flee, a fight will ensue. Both opponents approach each other with threat displays and twitching tail movements, and both may try to bite each other in the tail or flank.

The reproductive season begins about two months after the over-wintering period. The male approaches the female with a staggering gait, exhibiting threat behavior. He pushes repeatedly against her flank with his snout and also touches her cloacal region. Females unwilling to mate will react immediately with defensive biting. However, those ready to mate will remain in place. Initially the male applies a firm bite to the base of the female's tail then releases his grip in order to secure another firm bite just above her hind legs. The male then bends the posterior part of his body underneath that of the female. After mutual cloacal contact has been established, copulation follows and lasts about 20 to 30 seconds. During the breeding season a female remains willing to mate for about 4 to 5 days and so can copulate repeatedly with several males.

About 12 days after copulation the female begins to search for a damp area. Within 4 to 6 hours she excavates a hole into which she deposits from 2 to 12 eggs during the following 1½ hours. Once egg-laying has been completed the hole is filled in again, the female using her front legs to shovel sand from behind forward and into the hole. Finally the surface is smoothed over again with the snout so that hardly anything indicates that eggs have been buried there. The clutch is cautiously unearthed and the upper sides of the eggs are marked. The eggs are 10-12 mm long and have an average diameter of 5 to 7 mm. They should be transferred to a plastic bowl with damp sand and then to a Type I brood container at a temperature of 27 to 30° C. After 31 to 34 days the hatching lizard slits the tough egg shell with the help of an egg tooth located at the tip of the snout. At the time of hatching the young are approximately 60 mm long. They are best transferred to a semi-moist terrarium that offers many hiding places. Food in the form of small crickets, grasshoppers, fruitflies, and other small insects is taken from the first day on. The diet has to be vitamin-rich and varied. Regular UV radiation is essential. Sexual maturity is attained after 2 years.

The following species can be kept and bred under the same conditions:

Podarcis danfordi
Maximum size: 25 cm.
Distribution: Asia Minor.
Habitat: Rocky regions.

Clutch size: 15.
Incubation period: 43 days at 30° C.

Podarcis hispanica
Maximum size: 20 cm.
Distribution: Southern France, Spain, northwestern Africa.
Habitat: Rocky regions.
Clutch size: 12.
Incubation period and hatchling size: 63 days at 30° C; 7.4 cm.

Podarcis muralis—Wall Lizard
Maximum size: 18 cm.
Distribution: Western Balkans, Adriatic islands.
Habitat: Rocky regions, walls, gardens.
Clutch size: 8.
Incubation period and hatchling size: 45 days at 26-30° C; 3 cm.

Podarcis mellisellensis
Maximum size: 18 cm.
Distribution: Western Balkans and Adriatic islands.
Habitat: Rocky regions with bushes and shrubs.
Clutch size: 9.
Incubation period: 37 days at 27.5° C.

Podarcis taurica
Maximum size: 25 cm.
Distribution: Crimea, eastern and southern Balkans.
Habitat: Semi-damp lowlands and highlands.
Clutch size: 6.
Remarks: Sexual maturity in 2 years.

Lacerta armeniaca
Maximum size: 17.3 cm.
Distribution: Armenia to northwestern Azerbaijan.
Habitat: Rocky and stony regions.
Clutch size: 5.
Incubation period and hatchling size: 40 days at 30° C; 2.5 cm.
Remarks: Sexual maturity during the second year; in part parthenogenic reproduction.

Lacerta saxicola

Maximum size: 25 cm.
Distribution: Crimea, Caucasus, Asia Minor, Iran, Armenia, southern Turkmenia.
Habitat: Rocky regions.
Clutch size: 4.
Incubation period and hatchling size: 38 days at 20 to 30° C.
Remarks: In part parthenogenic reproduction.

Algyroides marchi

Maximum size: 15 cm.
Distribution: Southeastern Spain.
Habitat: Pine forests.
Clutch size: 3.
Incubation period and hatchling size: 34-38 days at 26-30° C; 6 cm.
Remarks: Over-wintering at 4 to 6° C recommended.

Ptychozoon lionotum—Flying Gecko
(Family Gekkonidae)

Description: Size to 16 cm. Dorsum with bark-like banded or striped pattern. Skin folds on sides of head, between toes, on arms and legs, and halfmoon-shaped skin folds margining both sides of the tail. Flanks with broad skin appendages. Males with 18-28 strongly developed pre-anal pores and two scale-like skin folds on each side of cloaca.
Distribution: Southeast Asia, Burma, Thailand, Island of Rami. Along tree trunks and behind tree bark.
Care: Food: Insects. Should be kept in a dry terrarium at air temperature from 24 to 30° C during the day and 18 to 22° C at night. Spray in the evening. Use sand or mixture of sand and peat moss as a substrate. The decoration can include climbing branches and tree or cork bark along the back wall of the terrarium. Succulents can be added. A small water bowl should be available.
Behavior and breeding: The bark-like coloration and skin-fold appendages of this species clearly indicate its habitat to be on tree trunks. This shows up also in captivity, where it can be found primarily on the tree or cork bark used as decoration in a terrarium. This is a nocturnal species that we keep together with other nocturnal gecko species in a terrarium 70 x

60 x 115 cm in size. Usually it pursues its prey on trees, with the body pressed closely against branches and tree trunks. Once prey has been spotted the gecko approaches it very slowly, stops briefly in front of it to observe it closely, and then seizes it with its jaws. The prey is slammed briefly against a branch or tree trunk—first to one side and then to the other—to stun it, and then it is swallowed. Subsequently the mouth and eyes are cleansed by licking.

If this gecko is up against the cork wall its body outline becomes totally lost in the background. Therefore, it does not come as a surprise to find that it tends to rely primarily upon its camouflage coloration when threatened. If it is about to be caught it will initially remain pressed firmly against the bark. Only when the flight distance (a few centimeters) has been exceeded will the animal attempt to escape with a sudden jump and then hide. When actually cornered there is a typical defensive reaction: the mouth is opened wide, the tail bent upward in a slight curve, and a groaning threat sound is given. Frequently this is followed by a sudden bite, the lizard remaining suspended from a finger or hand. The keeper should be faster and quickly grab the animal around the neck before it can bite. Although such a bite is harmless and can hardly be felt, it can cause jaw damage to the gecko, which may then be unable to feed properly.

Because of its cryptic, nocturnal behavior little is known about the reproductive mechanism. However, when a female becomes noticeably heavier and the 2 whitish eggs become clearly visible through the abdominal skin, the cork walls should be closely monitored (preferably the pieces of cork should be maintained so that they are removable). Since our females produce 3 successive clutches of 2 eggs each in 2-week intervals during one reproductive cycle, it is important that monitoring be maintained throughout this period.

A clutch usually consists of 2 hard-shelled eggs about 13 x 15 mm in length that are located either close to each other or may even be adhering to each other. When such a clutch is found it is best for the survival of the young geckos to remove the eggs together with the substrate they are adhering to and transfer them to a Type II brood container. The young will hatch after 32 days at a temperature of 19 to 30° C. They are 5.5 cm long and look just like their parents. Rearing them is usually without problems. They should be kept in a dry terrarium and be given a variable diet of insects supplemented with vitamins. After 5 months the young geckos will have doubled their length, and at that time they can be put together in the same terrarium with their parents. Sexual maturity is reached after 12 months.

> *Ptychozoon kuhli (15 cm; southeastern Asia, Indonesia; forest regions) can be kept and bred under identical conditions. Two hard-shelled eggs (14 x 10 mm) are laid, usually on tree bark. The young (6 cm) will hatch after 3 months at an incubation temperature of 24 to 26° C. Sexual maturity is reached after 12 months.*

Ptyodactylus hasselquistii—Fan-fingered Gecko
(Family Gekkonidae)

Description: Size to 14 cm. Dorsum beige to gray-brown, frequently with a brown stripe from the nose across the eyes. Reddish brown dorsal spots. Ventral area white. Head relatively large. Fan-like enlarged adhesive lamellae on tips of toes.

Distribution and habitat: North Africa to southwestern Asia. Desert regions, in rocky caves and along stone walls, in abandoned rock quarries or along banks of dry river beds.

Care: Food: Insects. Should be kept in a dry terrarium at air temperatures from 25 to 35° C during the day and 18 to 22° C at night. Give localized heating with some type of radiator and spray with water in the morning. Sand is usable as a substrate. There should be hiding and sunning places in the form of rock arrangements, plus the usual climbing branches. Succulents or similar plant species can be added, and a small water dish should be provided. This lizard can be overwintered at 10° C.

Behavior and breeding: If this species is to be caught in a terrarium it will display a similar defensive behavior as described for *P. lionotum*. Initially it will remain motionless on a branch, then nod its head several times and wave the slightly raised tail from side to side. When this gecko is attacked it emits a croaking defensive sound.

Mated pairs are particularly territorial during the breeding season. The male signals his territorial claim with a series of "tcha tcha" sounds. If another male invades his territory it will be severely attacked by the resident pair. For breeding purposes only one pair can be maintained in a small terrarium.

The female lays her eggs in 2- to 4-week intervals 4 to 5 times a year. She selects a suitable site along the sides or roofs of small caves where she can attach her clutch of 2 nearly round eggs (13-15 mm in diameter). The eggs, which are still soft when laid, are caught in the folded hind legs of the female, manipulated by her, pressed against each other, and then attached to the substrate. The clutch together with the section of substrate is then carefully removed and transferred to a Type II brood container. The incubation period lasts for 90 to 100 days at 18 to 25° C. The young (50 mm at birth) are reared in a dry terrarium on a diet of small insects.

Sceloporus jarrovi—Yarrow's Spiny Lizard, Yarrow's Fence Lizard

(Family Iguanidae)

Description: Size to 17 cm. Dorsum dark brown with blue iridescence, with a yellowish-margined blue-black neck band. Ventral area whitish, throat and sides of abdomen dark blue in males. Tail base of male thicker than in female.

Distribution and habitat: USA (southern Arizona) into Mexico. Gravel pits, rock piles, etc., with cacti and other succulents, rocks, logs, fence posts.

Care: Food: Insects, spiders, worms, blossoms. Should be kept in a dry terrarium at air temperatures from 22 to 35° C during the day and 15 to 22° C at night. Provide a water spray in the morning, localized heating with a radiator, and regular UV radiation. The substrate should consist of sand (half dry, the other half damp). Decorations can include climbing branches and rocky terraces. The back wall can be set up with pieces of tree or cork bark. Succulents may be planted. Provide a small water dish.

Behavior and breeding: This species is strongly territorial, as indeed are many members of this family. Therefore, aggressive action is common among males, particularly in a terrarium. If a male trespasses into the territory of another, the resident male initially just watches the trespasser. Then he proceeds to signal his territorial claims by nodding his head and repeated up and down motions of the anterior half of the body. If this does not deter the trespasser, the resident runs toward the other male, attempting to intimidate him with threat displays. For that he arches his back, flattens his body laterally, spreads his gular fold, and offers his deep blue flank to the opponent. If the latter does not flee immediately or signal his submissiveness by lowering his head and pressing his body to the substrate, a fight will commence. With a stalking gait and the body raised on all four legs, the opponents circle each other, continuously threatening, nodding their heads, and moving their tails from side to side. There are also attempts made to bite each other in the tail or abdomen. The fight ends when one of them flees.

During the courtship the male approaches a selected female, nodding in a species-specific rhythm. If she signals her mating readiness by raising her tail, the male butts her with his head several times in her flank and neck. Sometimes the male also licks the female. If the female still does not move, the male's head nodding increases and he lowers his head and bites firmly into her neck or shoulder region. He then moves the posterior part of his body several times over her back and posterior region. He maneuvers the posterior part of his body under hers and, after a short positioning period and cloacal contact, copulation takes place.

This species is live-bearing. When the female becomes heavier and movements become visible in her abdominal region, birth is imminent. Then, with her tail bent slightly upward, she will give birth to up to 13 young born at intervals of about 9 minutes each. At birth the young are 5.3 to 6 cm long and weigh 0.35 to 0.7 grams each. Initially they remain motionless inside the egg sac, but 5 to 10 seconds later they burst out with strong head and body movements and, with their eyelids blinking, they take their first breath of air. Shortly thereafter the young begin to lick their surroundings, then they start to nod their heads just like the parents and also move their torsos up and down.

The young are quite easy to raise. They should be housed in a dry terrarium and given a varied diet (small crickets, wax moth larvae, small grasshoppers, spiders, or similar items) with occasional multivitamin and mineral supplements. Regular UV radiation is required. Food is taken from their first day on. Sexual maturity is reached after about 5 months.

> *Sceloporus malachiticus* *(Southern Mexico to Panama)*
> *grows up to 20 cm. The male has a bright green body, while the*
> *throat, abdomen, and tail are turquoise; the female is colored less*
> *conspicuously. Habitat, care, and behavior approximate that of*
> **S. jarrovi.** *This species is also live-bearing. The pregnancy lasts*
> *3 to 4 months, and 3 to 5 young are born that are 5.3 to 6.3 cm*
> *long at birth. For details about raising the young, refer to* **S. jar-**
> **rovi;** *however, the young have to be maintained under somewhat*
> *moister conditions.*

Care and breeding of the following *Sceloporus* species conform largely to those discussed for the previous species:

Sceloporus cyanogenys—Blue Spiny Lizard
Maximum size: 36.2 cm.
Distribution: Texas, Mexico.
Habitat: Rocky areas.
Litter size: 18, live-bearing.

Sceloporus grammicus—Mesquite Lizard
Maximum size: 17.5 cm.
Distribution: Texas, Mexico.
Habitat: Arid regions with sparse plant growth.
Litter size and juvenile size: 16, live-bearing; 7 cm.

Sceloporus poinsetti—Crevice Spiny Lizard
Maximum size: 28.6 cm.
Distribution: Texas, New Mexico, Mexico.
Habitat: Rocky regions.
Litter size and juvenile size: 16, live-bearing; 7 cm.

Sceloporus graciosus—Sagebrush Lizard
Maximum size: 16 cm.
Distribution: Western USA.
Habitat: Brushland with sandy substrate.
Clutch size: 7.
Incubation period: 45 days; egg-laying.

Sceloporus magister—Desert Spiny Lizard
Maximum size: 30.5 cm.
Distribution: Southwestern USA, Mexico.
Habitat: Brushland, rocky areas.
Clutch size: 19.
Incubation period: 56-77 days; egg-laying.

Sceloporus occidentalis—Western Fence Lizard
Maximum size: 23.5 cm.
Distribution: Western USA.
Habitat: Rocky and forest regions.
Clutch size: 14.
Incubation period: 54-84 days; egg-laying.

Sceloporus undulatus—Eastern Fence Lizard
Maximum size: 19 cm.
Distribution: All warmer USA except far western area; northern Mexico.
Habitat: Forests, prairies, rocky regions.
Clutch size: 13.
Incubation period: 42-56 days at 25-32° C; egg-laying.

Xantusia vigilis (Xantusiidae)—Desert Night Lizard
Maximum size: 9 cm.
Distribution: Southwestern USA; northwestern Mexico.
Habitat: Desert to dry forests.
Litter size and juvenile size: 3, live-bearing; 5 cm.

Tarentola mauritanica—Wall Gecko
(Family Gekkonidae)
Description: Size to 16 cm. Dorsum light gray to black, with spiny or tuberculate scales. Ventral area white to yellow. Toes with adhesive lamellae on underside. Third and fourth toes on each foot with claws. Pupil vertical.
Distribution and habitat: Mediterranean region (Spain to Greece, Canary Islands, North Africa). In warm, dry, coastal lowlands on walls, ruins, rocks, and wood piles, as well as on exterior and interior building walls.
Care: Food: Insects, spiders. Should be kept in a dry terrarium at air temperatures from 22 to 35° C during the day and 15-22° C at night. Provide localized heating with a radiator of some type. Decorations can include the back wall covered

with cork bark, climbing branches, and rock structures. Plants to use include *Passiflora, Oleander* and similar types. A small water dish should be given. Can be over-wintered at 10 to 12° C.

Behavior and breeding: A well-known fact about geckos is their ability to scale up and down even vertical walls. What gives them this unique ability? Wall geckos, as well as many other species, have lamellar adhesive pads on their toes that consist of a multitude of tiny hair-like structures (as can be seen in electron microscopy), the setae. They enable geckos to hook into the tiniest of projections along even the smoothest surface.

If one threatens a wall gecko by reaching for it, it will raise the tail slightly and move its tip, curled up into an S-shape, from side to side. With this the gecko attempts to divert the attention of any potential attacker to the tip of the tail. If the tail is actually handled it will break off at a predetermined fracture line and continue to wiggle, while the gecko can then escape. The tail slowly regenerates. If a gecko is grabbed by the head it will emit a defense sound, a sort of chirping, and tries to twist its body out of the hand that is holding it.

During the breeding season, which extends from spring to summer in southern Europe, the males of this primarily nocturnal gecko are territorial. They occupy a territory, for instance part of a house wall, that is then defended against all other males. If one male trespasses the boundaries of another male's territory, an intense fight can occur.

According to observations in captivity, the eggs are laid during the late evening or night hours. The female digs a hole about 4 cm deep and 2 to 3 cm wide with her hind legs in damp sand. While doing this the toes are bent upward so that the adhesive lamellae do not come into contact with the loose substrate. Then the first egg is pushed out and caught with the folded hind legs, turned over, and dropped into the hole. After the second (sometimes even a third) egg has been laid the hole is filled in again, initially with the forelegs only, then also with the hind legs. Again, the adhesive lamellae are kept from coming into touch with the substrate. Frequently the female piles the sand somewhat high over the hole and then compacts the loose sand with her abdomen. According to natural habitat observations, the eggs are sometimes deposited in the cracks of walls or below rocks. The female can breed in 2- to 3-week intervals as often as 6 times per year. The eggs are 11-12 mm long and 9-10 mm in diameter. They have a hard shell, so they either can be left in the terrarium or can be taken out and transferred to a smaller Type II brood container. Dry sand and filter wool are suitable substrates.

With an average temperature of 26° C the incubation period lasts about 3 months. The young should be reared in a dry terrarium and be given a varied diet of fruitflies, small crickets, wax moth larvae, and small spiders. Occasional calcium and vitamin supplements are recommended.

Care and breeding (without over-wintering) of the following gecko species are essentially identical to the species discussed above:

Tarentola delalandii

Maximum size: 14 cm.
Distribution: Canary Islands.
Habitat: Rocky areas, buildings.
Clutch size: 2.
Incubation period and hatchling size: 116 days at 18-35° C; 4 cm.
Remarks: Breeding behavior and reproductive cycle correspond to *T. mauritanica* in our specimens received from southwestern Teneriffe. Only 1 egg is laid following each mating. This is buried in sand. The female guards the site for up to a week afterward and intruders are driven off. Should be kept in pairs only.

Gehyra mutilata—Stump-toed Gecko

Maximum size: 12 cm.
Distribution: Pacific Islands and coastline to Madagascar; introduced into Mexico and California.
Habitat: Agricultural land, buildings.
Clutch size: 2.
Incubation period and hatchling size: 45-48 days at 27° C; 5.7 cm.

Gehyra variegata

Maximum size: 15 cm.
Distribution: New Guinea, Australia, Oceania.

Habitat: Forests.
Clutch size: 1.
Incubation period and hatchling size: 84 days at 23-27° C; 6 cm.
Remarks: Sexually mature in 2 years.

Gekko gecko—Tokay Gecko
Maximum size: 30 cm.
Distribution: Southern and southeastern Asia.
Habitat: Forests, buildings.
Clutch size: 2.
Incubation period and hatchling size: 122 days at 28-33° C; 10 cm.

Gekko monarchus
Maximum size: 21 cm.
Distribution: India, Sunda Islands, Indo-Australian Archipelago, Philippines.
Habitat: Forests, buildings.
Clutch size: 2.
Incubation period: 100 days at 29-31° C.

Lepidodactylus lugubris
Maximum size: 8 cm.
Distribution: Northeastern Australia, southeastern Asia, Polynesia.
Habitat: Coastal forests.
Clutch size: 2.
Incubation period: 93 days at 22° C.
Remarks: Sexually mature in 8½ months. Parthenogenic reproduction to some extent.

Pachydactylus capensis
Maximum size: 20 cm.
Distribution: Southeastern Asia, Polynesia, South Africa.
Habitat: Rocky areas.
Clutch size: 2.
Incubation period and hatchling size: 48 days at 28° C; 4.8 cm.
Remarks: Eggs deposited in damp sand.

Rhacodactylus chahoua
Maximum size: 22 cm.
Distribution: New Caledonia.

Habitat: Forests.
Clutch size: 2.
Incubation period and hatchling size: 84-95 days at 27° C; 9.5 cm.

Sphaerodactylus cinereus—Ashy Gecko
Maximum size: 5 cm.
Distribution: Antilles to Florida Keys.
Habitat: Forests.
Clutch size: 2.
Incubation period and hatchling size: 82-86 days at 25° C; 3 cm.

Teratoscincus scincus—Wonder Gecko, Sand Gecko
(Family Gekkonidae)
Description: Size to 20 cm. Dorsum pink with light brown to blackish spots and bands. Dorsal surface of tail gray. Abdominal region white. Relatively large head. Flattened fingers and toes with small serrated scales. Skin thin, scales of head much smaller than body scales.
Distribution and habitat: Central and southwestern Asia. Desert regions with sparse plant cover.
Care: Food: Insects. Should be kept in a dry terrarium with air temperatures ranging from 30-42° C during the day and 15-25° C at night. Provide localized heating with a radiator. The substrate must include a damp sandy area (about half the surface) at least 25 cm deep; the other half should be dry sand. Decorations can include flat stones and rock structures. Plants (succulents) can be omitted. A small water bowl should be present.
Behavior and breeding: This species is a ground-dweller. It has a relatively small lung but is equipped with a thin skin with a vascular system which extends over the entire body surface. This facilitates oxygen exchange across the skin in addition to regular lung respiration. However, this arrangement has the disadvantage that these animals require a relatively high humidity, although they are inhabitants of a dry habitat. Therefore, in nature *T. scincus* spends most of the daylight hours in burrows it digs in damp soil, the burrows reaching down to about 80 cm. To conserve moisture a sand plug is used to close the entrance. This is a nocturnal species.

When threatened, *T. scincus* displays a characteristic defense behavior designed to scare off potential enemies. It raises its body high on all four legs, inflates the gular fold, and slowly waves the tail, which is bent into a S-shape. Rubbing together the tail scales, which are enlarged into horny plates, causes a rattling sound. Suddenly the animal lunges toward the enemy, simultaneously emitting a squeaking sound and trying to bite. After that the animal tries to flee back into a hiding place close by. This sort of behavior can be seen in juveniles only 2 weeks old.

Little is known about the reproductive behavior of this species, presumably due to its cryptic, nocturnal behavior. The female produces 2 eggs 2 or 3 times a year. The eggs are deposited in dry sand at the warmest possible site in a terrarium. They are 21-22 mm in size and have a very fragile shell. A fine paint brush should be used to uncover the eggs, which should be transferred to a Type II brood container with dry sand as the substrate. The eggs must not come into contact with water. The incubation period lasts 75 days at a temperature of 26-42° C. Upon hatching the young are 69-73 mm long. They display a well-defined juvenile coloration with a bright yellow dorsal region and jet-black crossbands. The young can easily be raised in a small dry terrarium, where they will soon chase after small crickets and similar sized insects.

The following geckos can be kept and bred under identical conditions:

Chondrodactylus angulifer
Maximum size: 15 cm.
Distribution: Southern and Southwestern Africa.
Habitat: Desert.
Clutch size: 1.
Incubation period: 57 days at 20-30° C.

Teratolepis fasciata
Maximum size: 9 cm.
Distribution: Western Pakistan.
Habitat: Desert.
Clutch size: 2.
Incubation period and hatchling size: 61 days at 28-30° C; 2 cm.

342

Remarks: Female guards clutch until young hatch.

Tiliqua gerrardi—Pink-tongued Skink
(Family Scincidae)

Description: Size to 45 cm. Dorsum silvery gray to brown with many dark brown crossbands. Ventral area gray to orange. Tongue pink.

Distribution and habitat: Eastern Australia. Semi-moist and dry forest regions. Among low vegetation on the ground.

Care: Food: Snails and slugs, insects, spiders, chicken liver, heart, and gizzards, beef heart, sometimes bananas. Should be kept in a semi-moist terrarium at air temperatures of 20-25° C during the day and 14-18° C at night. Spray with water daily and use localized floor heating. A peat moss and soil mixture about 10 cm deep and moss slabs make a good substrate. Decorations include climbing branches and pieces of dry roots, as well as *Scindapsus*, ferns, and similar plants. Supply a small water dish.

Behavior and breeding: In contrast to many other skinks, *T. gerrardi* becomes active during dusk and develops its full activities only during the night. At that time this species not only hunts along the ground but also climbs into low vegetation. Their claws and the long, prehensile tail are here particularly useful. If a snail has been spotted, the lizard runs toward it, constantly flicking his tongue and clearly following an olfactory gradient (scent trail) to the prey. Once the prey has been reached, the jaws are slammed around it. With the head held high, the prey is transported to a safe location and slammed sideways against the ground to crack the shell and then swallowed. Snails and slugs are the preferred food items. Therefore, it is important that a large supply be kept in storage in a freezer for the long winter months or a snail breeding colony be set up. If insects are fed, they are more eagerly taken when they have been rubbed against a snail.

In captivity, animals of the same sex frequently end up fighting among themselves. Therefore, it is important that the terrarium be sufficiently large with adequate escape and hiding facilities. During the mating season the female is frequently followed for days by a male that continuously tries to place a firm mating bite on her. Even a female initially willing to mate reacts defensively toward the male's courtship attempt, although eventually her resistance will wane. Finally, the male succeeds in securing a mating bite on the female's neck or below one of her forelegs. One of his forelegs strokes repeatedly over the female's back and his tail makes wave-like twitching movements. The female then raises her tail slightly so that the male can bring his tail underneath hers. Once cloacal contact has been established, copulation follows for about 4 minutes.

Pink-tongued skinks are live-bearing. The female produces up to 25 young after a pregnancy of about 5 months. At birth the young are approximately 10 cm long and possess the distinctive juvenile coloration of black crossbands against a light brown to reddish background color.

The young should be transferred to a separate semi-moist terrarium. Feeding commences on the third day after birth; small snails or pieces of snail meat are suitable first food. Sexual maturity occurs after 2½ years.

Tiliqua scincoides—Eastern Blue-tongued Skink
(Family Scincidae)

Description: Size to 50 cm. Dorsum gray to red-brown with dark brown to black crossbands. A dark band between the eyes and the nasal openings. Ventral area yellowish. The blue tongue is very obvious.

Distribution and habitat: Northern and eastern Australia, Tasmania. Semi-deserts, brushland, land under cultivation.

Care: Food: Insects, snails, newly born mice, lean chopped meat, strips of beef, canned dog and cat foods, fish, eggs, cottage cheese, vegetables, fruit. Should be kept in a dry terrarium at air temperatures from 25 to 35° C during the day and 15 to 22° C at night. Spray with water in the morning. Localized heating with a radiator of some type and floor heating plus regular UV radiation are necessary. Sand serves as a substrate. Decorations may include large stones and rocky structures as well as succulent plants, but the plants can be omitted. Provide a small water dish.

Behavior and breeding: When threatened, as when about to be caught by hand, this species exhibits a similar defensive behavior to that seen in *T. rugosus*. The animal raises its head slightly, opens the mouth wide, and displays the dark blue tongue. At the same time a hissing sound is emitted.

This is a ground-dwelling species that is active primarily during the day. If several animals are kept together in the same terrarium, aggression can occur—smaller and weaker animals are kept away from the food, and often they are attacked and bitten. Therefore, it is important that the terrarium is large enough and has sufficient hiding places as well as several feeding stations. If need be certain specimens may have to be fed individually by hand.

During the mating season the male follows the female, sometimes for hours, until he succeeds in establishing a firm mating bite in her neck region. About 4 months after copulation the female, which by then has become fairly heavy, will give birth to up to 25 young. Their first food consists of the afterbirth. Once they are transferred to a separate dry terrarium they will soon chase after crickets and similar insects as well as lick (just like their parents) on foods like sweet pudding, fruit, and similar items.

Tiliqua gigas (50 cm; New Guinea, Moluccan Islands, Indonesia; bushland) can be kept and bred under identical conditions. The female produces up to 10 young that are 15 cm long at birth.

Trachydosaurus rugosus—Shingleback, Pinecone Skink

(Family Scincidae)

Description: To 25 cm. Dorsum light brown to black, with or without yellow spots. Ventral area yellow-black. Scales rough, pinecone-like, especially prominent on the blunt tail that serves as a fat reservoir.

Distribution and habitat: Southern and southeastern Australia; desert regions.

Care: Food: Insects, snails, newly born mice, lean chopped meat, small pieces of beef, raw eggs, fruit, pudding, and similar items. Should be kept in a dry terrarium at air temperatures from 22 to 35° C during the day and 16 to 22° C at night. Provide localized heating with floor heating and a radiator, regular UV radiation, and sand as a substrate. The decoration can consist of rock structures and *Melaleuca, Callisternon, Grevillea,* or similar plants, but the plants can be omitted. A small water dish should be given.

Behavior and breeding: Shinglebacks have rather short legs relative to the rest of the body. Therefore, they are hardly in a position to flee quickly when threatened, and they have developed a special defense strategy. They bend their body into a semicircle, open the mouth widely, extend their dark blue tongue, and emit a threatening hissing sound. If this does not work the shingleback bites, and with its strong jaws it can easily cause bleeding wounds.

During the courtship the male follows the female for several days. He repeatedly licks her tail and cloacal region as well as her sides and head. Then the male crawls over the female's back and attempts to secure a firm mating bite on her forelegs or sides of her head. This is not always easy, because the female keeps getting away. Finally he secures a firm bite to a side of the head (cheek) and bends his body under that of the female, moves over to the side opposite to his bite, reaches with one hind leg for the base of the female's tail, and then establishes cloacal contact.

Just like many other Australian skinks, shinglebacks are live-bearing. One female produces 2 to 3 young per year that are approximately 14 cm long at birth. Their first food is the afterbirth. From their third day on they will lick on sweet fruit and puddings and start to chase after slow ground insects.

Varanus gilleni—Pygmy Mulga Monitor
(Family Varanidae)

Description: Size to 34 cm. Dorsum gray-brown with brown to reddish crossbands. Claws sharp, tail prehensile.

Distribution and habitat: Central Australia. Desert regions with mulga, eucalyptus, and she-oak trees. Found under loose bark.

Care: Food: Insects. Should be kept in a dry terrarium at air temperatures from 25 to 30° C during the day and slightly lower temperatures at night. Provide localized heating with floor heat and a radiator, plus regular UV radiation. The substrate can consist of a layer of sand at least 10 cm deep (half of it damp and the other half dry). Decorations can include rocks, hollow logs or branches, and climbing branches. A small water dish should be available.

Behavior and breeding: This monitor is active primarily during the day. If several males are kept together with one female there are frequent fights, particularly during the breeding season. If two males come across each other they initially try to intimidate each other by enlarging their body profile. They turn toward each other, inflate their gular folds, flatten their bodies dorso-ventrally, and show their enlarged body profiles from positions oblique to that of the opponent. This is followed by a characteristic species-specific ritualistic fight. Both opponents move toward each other signalling threats. Suddenly they thrust themselves at each other, wrapping their legs around the rump of the opponent in order to dislodge him in this belly-to-belly position by turning him over on his back. Both head and tail are used in these maneuvers, acting as balancing stabilizers. If neither of the two succeeds in gaining an advantage over the other, both will inflate and turn their body so that one side of the rump touches one side of the opponent's rump. While doing that, both maintain their grip with their claws on the opponent. Then, while exhaling they rotate jointly around their longitudinal axis. These procedures are repeated at intervals until one of the opponents succeeds in pressing the other to the ground, bringing his front and hind legs in front of those of the opponent, and forcing his throat region to the ground. Then both animals let go. Frequently the dominant animal will then bite the loser in the cloacal region. If the loser then closes his eyes and presses head, rump, and tail close to the ground, he will be protected against further attacks. However, he is then frequently courted just like a female.

During the courtship a male keeps following a particular female, executing lateral zig-zag movements with his head and neck region. The intensity, frequency and duration of these movements increase with the degree of excitement of the male.

He keeps pushing his head laterally against the cloacal region of the female and so attempts to stimulate the female into mating. If the female is unwilling she will react defensively, biting the male and then fleeing. However, if the male succeeds in establishing a mating bite to the neck of a willing female, he reaches around her pelvic girdle with his hind legs and maneuvers his tail under hers. Copulation takes place for about 30 minutes.

Approximately 3 weeks later the female has become noticeably heavier. She then starts to dig a tunnel-like nest burrow about 40 cm long in the damp section of the sandy substrate. At the end of this tunnel she deposits 3 white parchment-like eggs that are 30 x 14 mm in size. The entrance to the tunnel is subsequently sealed off and the loose sand is smoothed over. The eggs should be removed carefully and their upper side properly marked. They are then transferred into a Type I brood container.

At 29-30° C the 11-cm young hatch after 87 to 95 days. They should be transferred into a separate dry terrarium and given a diet of crickets, grasshoppers, wax moth larvae, and similar items. Regular UV radiation is important.

The following monitors can be kept and bred under identical conditions (terrarium dimensions and food item size varying according to size of the species):

Varanus brevicauda
Maximum size: 20 cm.
Distribution: Western Australia.
Habitat: Dry regions.
Clutch size: 3.
Incubation period: 70-84 days at 18-25° C.

Varanus exanthematicus
Maximum size: 200 cm.
Distribution: Africa south of the Sahara.
Habitat: Dry regions.
Clutch size: 40.
Incubation period and hatchling size: 170-176 days at 26.7-35° C; 8 cm.
Remarks: Large terrarium required.

Varanus storri
Maximum size: 44 cm.
Distribution: Northern Australia.
Habitat: Dry forest.
Clutch size: 4.
Hatchling size: 12.3 cm.
Remarks: Adults should be kept in pairs only.

Varanus timorensis—Spotted Tree Monitor
(Family Varanidae)
Description: Size to 60 cm. Dorsum gray to black with white or yellowish spots. Markings highly variable.

Distribution and habitat: Northern Australia, Timor, New Guinea. Semi-moist forest regions; on trees along water.

Care: Food: Insects, newly born mice, strips of lean beef, fish. Should be kept in a semi-moist terrarium at air temperatures ranging from 25 to 30° C during the day and 20 to 22° C at night. Provide localized heating with a radiator plus regular UV radiation. The substrate should consist of leafy soil or a mixture of sand and peat moss. Decorations should include climbing branches. A water section with a container for bathing should be provided.

Behavior and breeding: In its search for food this monitor, as well as the other members of this family, orients itself largely on the basis of odors. It can detect the presence of prey through repeated tongue flicking. Such prey is then approached slowly, and suddenly the monitor lunges forward to seize the prey with its jaws. By slamming its head and jaws (holding the prey) sideways against the ground, the prey is stunned or killed. With the head held high and several gulping motions, the prey is then swallowed; often this involves snake-like movements of the long neck typical of monitors.

The markings of spotted tree monitors have made them well adapted to their environment. In the event of imminent danger they press their body flat against the branch or tree trunk so that their body outline becomes fused with the bark. Alternatively, they may also attempt to hide by moving to the opposite side of the tree trunk so they will not be noticed by an enemy.

Little is known about the social and reproductive behavior of this species. Allegedly, fighting is common among males when they are kept together in captivity. Therefore, subordinate males should be moved into separate accommodations.

This species is one of the few smaller monitors that have been bred in captivity. About 40 days after copulation the female deposits 9 eggs under the roots of plants somewhere in the damp substrate. It is advisable to remove the eggs from the terrarium for better control. Their upper side should be properly marked, and then the eggs are placed in a plastic container with a damp mixture of peat moss and sand. Incubation should be done in a Type I brood container. The 15.3 cm young will hatch after about 140 days at 28-31° C. Upon hatching they are transferred to a semi-moist terrarium. Initially they will derive nourishment by absorbing their own yolk sac. Actual feeding commences from about the third day on. Insects, small strips of beef, and similar items are taken.

Varanus mertensi (1.3 m; Australia, along water with heavy vegetation) can also be kept and bred under the same conditions, but in a very large terrarium. The female lays up to 14 eggs. The incubation period lasts 182-217 days at 29-30° C. The newly hatched young will start feeding from about the fifth day on. The diet should consist of fish, insects, and mice.

Varanus rudicollis (1.6 m, southeastern Asia, forests). Clutch size up to 14 eggs. Incubation period 180-184 days at 29° C. Juvenile size 25 cm.; for further details see V. mertensi.

REPTILES
Snakes (Serpentes)

Boa constrictor—**Common Boa**
(Family Boidae)
Description: Size to 4 m. Dorsum yellowish, reddish, or brownish with complicated light and dark pattern. Dark band on face from tip of snout to neck. Dark bands on top of head from in front of eyes to neck. Ventral area gray with dark spots.

Distribution and habitat: Mexico south to Argentina, Lesser Antilles. Forests and brushland in the proximity of water.

Care: Food: Small mammals, birds. Should be kept in a dry terrarium with air temperatures from 25-32° C during the day and 20-22° C at night. Localized heating with a radiator plus floor heating and regular UV radiation should be provided. For a substrate use sand or a mixture of peat moss and sand. Climbing branches should be provided, as should a water basin.

Behavior and breeding: This is essentially a nocturnal snake that frequents trees as well as the ground. It is the giant snake most frequently kept and bred in captivity.

In captivity the mating season extends over about 4 weeks. During the courtship period the male keeps following the female. With his head raised, the male keeps crawling over the female's back, caressing her with his erected anal spurs. Presumably this stimulates the female into mating.

Boa constrictor, as all other boas, is live-bearing. About 15-17 months after copulation, the female gives birth to about 60 young that are 35 to 50 cm long. They should be transferred to a separate dry terrarium. The first molting occurs after 5 to 21 days, and the first food (small mice) is taken about 9 to 15 days later. Sexual maturity is attained after approximately 3 years.

Candoia carinata (1.5 m; Solomon Islands) can be bred under the same conditions. Seven months after copulation the female produces up to 27 young that are 22-24 cm long and weigh 4.2 to 4.6 grams. Feeding (newly born mice) will commence after the first molt.

Coluber constrictor—Racer

(Family Colubridae)

Description: Size to 2 m. Dorsum olive-brown to black or bluish (occasionally with white dots or mottling). Ventrally black to greenish yellow or white. Many subspecies that vary greatly in color. Young with reddish brown dorsal blotches.

Distribution and habitat: Southern Canada south to Guatemala. Dry and wet regions; meadows, forests, cultivated land up to 2000 m elevation.

Care: Food: Small mammals, birds, lizards, snakes, sometimes insects, frogs. Should be kept in a semi-moist terrarium at air temperatures ranging from 20-29° C during the day and 18-20° C at night. Provide floor heating. A sand and forest soil mixture will do as a substrate. Decorations can include roots and climbing branches. Provide a large water bowl.

Behavior and breeding: This snake is active primarily during the day. Its mating season occurs in spring. The male follows the female for several hours, then he places his body next to hers, executing repeated sinuous body movements while his body moves intermittently forward and back again. Sometimes the male turns away briefly, encircling the female and then repeating his courtship attempts. If the female signals her willingness to mate she raises her tail, and after some searching the male presses his cloaca against hers and applies a mating bite to her neck. Copulation occurs after the hemipenis has been inserted. The female signals the end of her mating willingness through an increased restlessness, and she crawls forward, pulling the male along. Finally the male lets go and both animals turn away from each other.

About two months after copulation the female lays up to 42 eggs at a secluded site. After they have been properly marked on the upper side they are then transferred to a Type I brood container. The 25-cm young will hatch after 51 to 63 days at 24 to 28° C. Upon hatching the young display the typical juvenile coloration of a gray back with gray, brown, or reddish spots and dark spots on the belly. They should be maintained in a separate semi-moist terrarium and be given a diet (after the first molt) of small mice.

Keeping and breeding the following species correspond to that of the previous species:

Drymarchon corais—Indigo Snake
Maximum size: 3 m.
Distribution: Southeastern USA to Argentina.
Habitat: Lowlands in dry to moist regions.
Clutch size: 6.
Incubation period and hatchling size: 75-77 days at 28° C; 60 cm.

Masticophis flagellum—Coachwhip, Whipsnake
Maximum size: 2.6 m.
Distribution: Eastern USA to Mexico.
Habitat: Dry and wet regions.
Clutch size: 20.
Incubation period and hatchling size: 70-79 days at 25-30° C; 35 cm.

Dasypeltis scabra—African Egg-eating Snake
(Family Colubridae)
Description: Size to 80 cm. Dorsum gray, yellow, or brown with longitudinal rows of black dots. Two black spots, meeting at acute angle, on top of head. Ventral area white to yellowish with or without dark spots.
Distribution and habitat: Southern Arabian Peninsula and all but northern Africa. Dry brush and forest regions.
Care: Food: Bird eggs, including those of chickens, pigeons, quail and finches. Should be kept in a dry terrarium at air temperatures from 22-28° C during the day and 15-20° C at night. Provide localized floor heating and regular UV radiation. Use sand as a substrate (half damp, half dry). Climbing branches and rocks should be included, as may sansevierias or similar plants. Provide a water bowl.
Behavior and breeding: This species is active mainly during the hours of dusk and total darkness. It has (for snakes) extraordinary nutritional requirements: it feeds exclusively on eggs. How is it possible for a snake to reach the nourishing contents of an egg without hurting itself? Although egg-eating snakes have a reduced number of teeth, they also have extremely elongated projections from the lower sides of the 24th to 30th vertebrae. These actually penetrate into the

esophagus and are covered with a hard dentine-like layer. The snake locates eggs by flicking its tongue and thus utilizes an extremely keen sense of smell and taste combined. Once an egg has been found, the snake pushes it against some object, opens its mouth to the extreme, and with alternate movements of the upper and lower jaws begins to swallow the egg. Continuous muscular movements of the esophagus then push the egg toward the stomach and slice the egg shell open by pushing it against the spinous projections—the so-called egg saw. The liquid egg contents proceed toward the stomach, while the spinous projections prevent large and potentially dangerous shell fragments from following; these are ultimately regurgitated. When threatened this species will display a rather characteristic defensive behavior. The snake curls up, rubbing the keeled body scales along each other and causing a rattling sound. When danger is imminent the body is also inflated, the head and neck region is raised, the upper body is pulled back into an S-shape, and then the snake suddenly lunges toward the enemy with the mouth wide open.

Dasypeltis scabra has been bred repeatedly in captivity. The female deposits 12 to 15 eggs at a protected site in the terrarium. The eggs are 34 to 40 mm long, 16 to 18 mm in diameter, and weigh 6.6 to 8.6 grams. After the upper side has been properly marked the eggs are then transferred into plastic bowls with damp sand. These in turn are placed inside a Type I brood container. The incubation period is 54-63 days at 28-30° C. At birth the young are on average 27.9 cm long and weigh 5.6 grams. They are transferred to a dry terrarium, where they should be given a diet of small eggs about the size of sparrow eggs.

Elaphe guttata—Corn Snake
(Family Colubridae)
Description: Size to over 180 cm. Dorsum pale brown to brick red with large brownish to brick red spots surrounded by black margin. V-shaped spots from the back of the head onto the neck. Tail striped below.
Distribution and habitat: Most of the eastern and central USA west to the Great Plains and then south into northeastern Mexico. Pine forests, fields and paddocks, abandoned orchards, rock piles, small rodent burrows. Often around abandoned sheds.

Care: Food: Small mammals, birds. Should be kept in a dry terrarium at air temperatures ranging from 22 to 28° C during the day and 18 to 20° C at night. Use mixtures of sand and forest soil or sand and peat moss as the substrate (half of it damp, the other half dry). Decorations should include climbing branches and rocks, and a water bowl should be given. Over-wintering at 5 to 15° C is possible.

Behavior and breeding: During spring, corn snakes are active throughout the day, but during summer they tend to become more nocturnal. If an attempt is made to pick up a corn snake, it will assume a defensive posture. The anterior part of the body is pulled back into an S-shape and the snake then lunges toward the presumed enemy and bites. They usually become tame after a while in captivity.

This attractive colubrid usually mates in spring. The male tends to follow a selected female for hours then crawls repeatedly over her back, executing wave-like body movements. With a sudden movement of the posterior part of his body, the male manages to entwine the female's tail and so establish cloacal contact. During the subsequent copulation, lasting from 9 to 14 minutes, both partners raise and lower their tails, usually in a synchronized fashion.

Approximately 3 months later the female deposits 12 to 24 eggs (which tend to adhere to each other) in a protected location. The eggs are 2.5-3.9 cm x 1.7-1.8 cm and weigh about 5 to 8 grams. Their upper sides should be marked properly, and then they are transferred to plastic bowls filled with damp peat moss. The eggs are incubated in a Type I brood container at 25° C. The young will hatch after about 55 days and have a length of 20-24 cm. They are moved to a dry terrarium. After the first molt they will start to feed (after about 10 days) on very young mice. Sexual maturity is attained after 2 years.

Care and breeding of the following colubrids largely conform to that of the previous species:

Elaphe carinata
Maximum size: 2 m.
Distribution: China, Taiwan.
Habitat: Brushlands and forest regions.
Clutch size: 7.

Incubation period and hatchling size: 42-52 days at 28° C; 30 cm.
Remarks: Will feed on other snakes. Over-wintering is advisable.

Elaphe dione
Maximum size: 1 m.
Distribution: Southeastern Europe to Korea and China.
Habitat: Wet and dry regions.
Clutch size: 16.
Incubation period and hatchling size: 14-36 days at 22 to 28° C; 20 cm.
Remarks: Will feed on lizards, amphibians, and fishes. Over-wintering is possible.

Elaphe longissima
Maximum size: 2 m.
Distribution: Northeastern Spain across central and eastern Europe to Asia Minor and northern Iran.
Habitat: Outer forest areas, sunny vegetation-covered rocky slopes, brushland around fields and paddocks.
Clutch size: 18.
Incubation period: 60 days at 24-26°C.
Remarks: Over-wintering desirable.

Elaphe obsoleta—Rat Snake
Maximum size: 2.6 m.
Distribution: Eastern and central USA south to the Mexican border.
Habitat: Wet and dry regions.
Clutch size: 44.
Incubation period and hatchling size: 74 days at 27° C; 32 cm.
Remarks: Will also feed on amphibians; over-wintering desirable.

Elaphe quatuorlineata
Maximum size: 2.1 m.
Distribution: Southeastern Europe and western Asia.
Habitat: Brushy dry and rocky slopes to swampy banks and shorelines.
Clutch size: 16.
Incubation period and hatchling size: 30-60 days at 27° C; 35 cm.

Remarks: Over-wintering desirable.

Elaphe schrenki
Maximum size: 1.7 m.
Distribution: Eastern USSR, Korea, northeastern China.
Habitat: Forest and brushlands.
Clutch size: 30.
Incubation period and hatchling size: 48-50 days at 25-28° C;
30 cm.
Remarks: Over-wintering desirable.

Elaphe situla
Maximum size: 1 m.
Distribution: Southern Italy, Sicily, Malta, Balkans, Crete,
Caucasus, Crimea.
Habitat: Dry regions with sparse vegetation.
Clutch size: 5.
Incubation period and hatchling size: 60-70 days at 24-28° C;
31 cm.
Remarks: Over-wintering possible.

Elaphe subocularis—Trans-Pecos Rat Snake
Maximum size: 1.7 m.
Distribution: Southwestern USA to northern Mexico.
Habitat: Desert.
Clutch size: 7.
Incubation period and hatchling size: 80-90 days at 24-30° C;
28 cm.
Remarks: Sexual maturity in 2-3 years; will also feed on
lizards.

Elaphe vulpina—Fox Snake
Maximum size: 1.8 m.
Distribution: Northcentral USA.
Habitat: Proximity of water.
Clutch size: 29.
Incubation period and hatchling size: 56 days at 28° C; 40 cm.
Remarks: Over-wintering desirable.

Eryx johnii—Indian Sand Boa
(Boidae)
Description: Size to 100 cm. Dorsum yellow to dark brown.

Head and tip of tail similar in shape, hard to distinguish. Upper jaw projects over lower jaw. Median furrow along back.

Distribution and habitat: Pakistan and India. Dry regions, steppes and open forests.

Care: Food: Small mammals, birds. Should be kept in a dry terrarium at air temperatures ranging from 25-35° C during the day and 20-22° C at night. Supply localized floor heating and regular UV radiation. Sand is a good substrate (it must be at least 25 cm deep). Decorate with branches and flat stones. Give a water basin.

Behavior and breeding: Sand boas spend the day buried in desert sand. Only after the onset of dusk do they emerge and start hunting for prey. During the breeding season neither male nor female will feed. The male moves through the terrarium continuously flicking its tongue, presumably tracking odor trails of females. If it happens to come across a partially buried female, the male will bury his head slightly in the sand and try to raise the female's tail. He then crawls over the body of the female, making massaging movements along her back. As soon as the male's cloacal region touches the female, his anal spurs become erect and start to move rapidly up and down while scratching over the back of the female as he moves along her body. In this manner the male continues to court the female, crawling over her in both directions. Finally, he slides underneath the female, turning himself over onto his back and pressing his tail firmly against that of the female. After a brief search for a firm hold, copulation takes place. The posterior third of the male's body executes wave-like movements while the tail makes twitching up-and-down motions.

About 4 months later the female releases up to 16 young that are 27-28 cm long and weigh 20-22 grams; they still possess a 2-cm umbilical cord that dries up during the following 10-12 days. The young snakes should be kept in separate dry terraria. The first molt takes place after about 2 weeks. Then they will start to feed on very young mice. Young that continue to refuse food for some time should be force-fed.

The following *Eryx* species can be bred under identical conditions:

Eryx colubrinus
Maximum size: 70 cm.
Distribution: Northern and eastern Africa, Arabian Peninsula.
Habitat: Dry regions.
Litter size: 15.
Juvenile size: 19.8 cm.

Eryx jaculus
Maximum size: 80 cm.
Distribution: Southeastern Europe, southwestern Asia, northern Africa.
Habitat: Steppes, deserts with sparse vegetation.
Litter size: 15.
Juvenile size: 12 cm.
Remarks: Over-winter at 5 to 10° C.

Eryx miliaris
Maximum size: 95 cm.
Distribution: Kazakhstan to western Mongolia, Afghanistan.
Habitat: Semi-desert.
Litter size: 10.
Juvenile size: 13 cm.
Remarks: Over-winter as *E. jaculus*.

Eryx tataricus
Maximum size: 98 cm.
Distribution: Kazakhstan, central Asia, western China, Iran, Afghanistan, northwestern Pakistan.
Habitat: Dry regions.
Litter size: 20.
Juvenile size: 16 cm.
Remarks: Over-winter as *E. jaculus*.

Lampropeltis getulus—Common Kingsnake, Chainsnake
(Family Colubridae)
Description: Size to 180 cm. Dorsum black to brownish or greenish, with or without more or less distinct chain pattern

made up of white to yellowish crossbands, or small white to yellow spots, or whitish dorsal bands. Ventral area spotted or striped. Many subspecies that vary greatly in pattern.

Distribution and habitat: Widely distributed over the southern three-quarters of the USA and into Mexico. Damp pine forests with sandy plateaus, densely overgrown river and lake banks or shorelines, meadows, prairies, and near-desert conditions depending on subspecies.

Care: Food: Small mammals, birds, snakes, lizards. Should be kept in a semi-moist terrarium with air temperatures from 25 to 30° C during the day and 20 to 22° C at night. Provide localized floor heating. The substrate must consist of at least 10 cm of a mixture of leafy top soil and sand. Decorations can include rocks and pieces of bark. Give a water bowl. Can be over-wintered at 5 to 15° C, but this is not a mandatory prerequisite for breeding.

Behavior and breeding: When threatened, this species displays a characteristic defensive behavior. It curls up, rattles its tail, points its S-shaped anterior body in the direction of the enemy, and suddenly lunges forward. Immediately in front of the enemy the snake opens its mouth just wide enough for the front teeth to make contact with the enemy.

The breeding season for *Lampropeltis getulus* is in spring. During the courtship a male follows a selected female for about 20 minutes, continuously biting her. Then a mating bite is applied to the anterior section of the female's back and the male climbs onto her back, with his body making wave-like movements. Both partners vibrate the tips of their tails. After approximately an hour the male will loosen his grip on the female and move away, while the female curls up and remains motionless. Shortly thereafter the male will grab the female again, making the characteristic wave-like motions. He pushes the female and bites her in the flank. Finally he presses his cloacal region against hers and inserts the hemipenis. Courtship and mating last for about 4 hours.

Approximately 75 days after copulation the female produces 5 to 17 eggs that are 42-50 mm long with a diameter of 20-22 mm. The upper side of the eggs should be properly marked and the eggs transferred into a plastic container filled with a damp peat moss and sand mixture or peat moss alone, which is then incubated in a Type I brood container. The young will hatch after 60-70 days at 22-30° C; at that stage

360

they are 24-29 cm long. They are moved to a semi-moist terrarium. The first molt occurs after 10-14 days, and soon thereafter the young snakes start to feed on very young mice. CAUTION: Juvenile snakes (and adults) sometimes display cannibalistic tendencies! Sexual maturity is reached in 2½ years.

The following *Lampropeltis* species can be kept and bred under similar conditions:

Lampropeltis mexicana—Gray-banded Kingsnake
Maximum size: 120 cm.
Distribution: Southwestern Texas into eastern Mexico.
Habitat: Desert and prairie.
Clutch size: 6.
Incubation period and hatchling size: 70-91 days at 22-28° C; 22 cm.
Remarks: Over-winter.

Lampropeltis pyromelana—Sonora Mountain Kingsnake
Maximum size: 107 cm.
Distribution: Utah, Arizona, and New Mexico south into Mexico.
Habitat: Mountain regions, pine forests.
Clutch size: 6.
Incubation period and hatchling size: 71-72 days at 20-33° C; 24 cm.
Remarks: Over-winter.

Lampropeltis triangulum—Milk Snake, Scarlet Kingsnake, Tricolor Kingsnake
Maximum size: 130 cm.
Distribution: Eastern two-thirds of USA to Central America and northwestern South America. Many subspecies.
Habitat: Wide range of habits depending on subspecies, from deserts and pine barrens to mountains and bayou lowlands.
Clutch size: 24.
Incubation period and hatchling size: 49-63 days at 24-28° C; 18 cm.
Remarks: Over-winter.

Lampropeltis zonata—California Mountain Kingsnake
Maximum size: 95 cm.

Distribution: Mountain ranges of California and adjacent areas along western coastal USA.

Habitat: Pine and oak forests.

Clutch size: 4.

Incubation period and hatchling size: 66-85 days at 25-28° C; 23 cm.

Remarks: Over-winter.

Liasis childreni—Children's Python
(Family Boidae)

Description: Size to 180 cm. Dorsum brown to light gray with irregular dark brown patches. Ventral area white to yellowish. Dark brown stripe behind each eye.

Distribution and habitat: All but southern Australia. Coastal rain forests to desert regions. Underneath rock slabs in crevices and termite mounds.

Care: Food: Small mammals, birds, lizards. Should be kept in either a dry or semi-moist terrarium at air temperatures ranging from 25-35° C during the day and about 22° C at night. Sand, a peat moss and sand mixture, or layers of fresh moss can be used as substrate. Decorations can include rock structures and climbing branches. A water bowl for bathing is required.

Behavior and breeding: This species of python remains relatively small. It is essentially nocturnal in its activities, but in captivity this snake will also pursue prey during daylight hours. As soon as prey such as a mouse is introduced into the terrarium, the snake will raise its head and neck and start flicking its tongue. With the aid of a remarkable olfactory sense and infrared-sensitive lip grooves along the lower jaw, the snake can determine the type and location of the prey. If the snake is hungry it will follow the mouse slowly, continuously flicking its tongue. Then, when close to the prey the snake suddenly retracts its head and neck region into an S-shape, lunging at the prey and seizing it with wide-open mouth. Within seconds the snake coils itself around the mouse, suffocating it. When the prey has succumbed the snake gradually unwinds and tests the prey with its tongue. As soon as the head of the mouse has been found the snake will swallow it head-first. For that the upper jaws and then each side of the lower jaw are alternately moved over the prey to push the mouse into the enormously expandable mouth.

As in many other pythons, mating activities can also be triggered in *L. childreni* by initially keeping males and females apart and lowering the temperature to 18-22° C, as well as by instituting reduced daylight hours. Then, when males and females are brought together and the temperatures have been increased again, mating will occur after some initial ritualistic fighting among the males.

About 5-6 months after copulation the female will lay up to 12 eggs. She coils her body around the eggs and guards them. At temperatures from 27 to 31° C and a relative humidity of approximately 100% (the clutch should be misted with water or transferred to a Type I brood container for incubation) the young will hatch after about 65 days. They are on the average 30 cm and weigh 7 grams. The young are best transferred to a separate rearing terrarium. Following the first molt at the age of about one week, food is taken, usually in the form of very young mice. Sexual maturity is attained after 18 months.

Liasis mackloti (including *fuscus*) (3 m; Timor, New Guinea, northern Australia; open forests in proximity of water) can be bred under identical conditions. The female guards her eggs by coiling around the clutch. Up to 17 eggs (55 x 60 mm). Incubation period 67-71 days at 28-30.5° C. The young are 40-48 cm long and weigh 26-32.5 grams. For rearing details see *L. childreni*. The young will start feeding on newborn mice within 3 to 5 days.

Lichanura trivirgata—Rosy Boa
(Family Boidae)

Description: Size to 100 cm. Dorsum brown to beige or blue-gray, with three irregularly serrated brown to reddish longitudinal bands. Ventrally whitish with gray spots and blotches. Color quite variable.

Distribution and habitat: Southwestern USA to northwestern Mexico. Desert flats and rocky areas with brushy vegetation.

Care: Food: Small mammals, birds. Should be kept in a semi-moist terrarium at air temperatures ranging from 20 to 26° C during the day and 18-20° C at night. Provide localized floor heating. A mixture of sand and leafy forest soil is good as a substrate. Decorations can include flat rocks or pieces of cork bark, climbing branches, and a water bowl. This snake can be over-wintered at 10 to 15° C.

Behavior and breeding: This boa is primarily a ground-dweller. It has developed a defense strategy identical to that of the African ball python: when threatened, the rosy boa coils up into a ball and retracts its head into the center of the coils. This then provides the smallest possible body and surface area for the enemy to attack.

During the courtship a male approaches a female, continuously flicking his tongue. The male keeps sliding alongside the body of the female, running his flicking tongue over her and moving up to her head. He then places the hind part of his body over hers and moves his tail with the anal spurs erected over her tail section. If the female is ready to mate she will roll over slightly to one side, elevating her tail somewhat. This enables the male to establish cloacal contact and insert the hemipenis. This is followed by copulation, which lasts for about 20 minutes.

The rosy boa is a live-bearer, as indeed are all boas. Approximately 130 days after the mating the female produces 4-12 juveniles that are about 30 cm long at birth. They should be accommodated in a separate dry terrarium and be given a diet of small mice. Sexual maturity is attained after about 2 years.

Care and breeding of the following snakes conform largely to that of the rosy boa:

Epicrates cenchria (Boidae)—**Rainbow Boa**
Maximum size: 200 cm.
Distribution: Costa Rica to Argentina.
Habitat: Rocky areas, forests, plantations.
Reproduction and juvenile size: Live-bearing; gestation 5 months; 55 cm.
Remarks: Adults should be kept at 26-32° C during the day and 22-25° C at night.

Epicrates striatus (Boidae)—**Antillean Boa**
Maximum size: 300 cm.
Distribution: Hispaniola, Bahamas.
Habitat: Bushland, mangrove regions.

Reproduction: Live-bearing; 51 young.
Remarks: Over-winter; will also feed on lizards.

Corallus enydris (Boidae)—Garden Tree Boa
Maximum size: 250 cm.
Distribution: Southern Central America to northern South America.
Habitat: Rain forests.
Reproduction: Live-bearing; 30 young.
Remarks: Adults should be kept at 25-33° C during the day and 20-24° C at night in a moist terrarium; plants can include *Hoya, Philodendron,* or similar species; use peat moss as a substrate; spray repeatedly.

Trachyboa boulengeri (Boidae)—Spiny Boa
Maximum size: 40 cm.
Distribution: Ecuador, Colombia, Panama.
Habitat: Rain forests.
Reproduction: Live-bearing; gestation 10 months; 6 young, 12.5 cm long.
Remarks: Care as above; will also feed on fishes, amphibians, and insects.

Chondropython viridis (Boidae)—Green Tree Python
Maximum size: 180 cm.
Distribution: New Guinea, northern Australia.
Habitat: Rain forests.
Clutch size: 12.
Incubation period and hatchling size: 48-51 days at 22-23° C; 30 cm.
Remarks: Care as above; female guards eggs.

Python curtus (Boidae)—Blood Python
Maximum size: 300 cm.
Distribution: Indonesia, Borneo, Malaysia.
Habitat: Rain forests.
Clutch size: 16.
Incubation period and hatchling size: 70-75 days at 30-33° C; 39 cm.
Remarks: Care as above; female guards eggs.

Gonyosoma oxycephalum (Colubridae)—**Pale-tailed Rat Snake, Red-tailed Racer**
Maximum size: 230 cm.
Distribution: Southeastern Asia, Indonesia, Philippines.
Habitat: Rain forests.
Clutch size: 8.
Incubation period and hatchling size: 98-119 days at 28° C; 46 cm.
Remarks: Care as above; sexual maturity in 4 years.

Natrix natrix—Ringed Snake, Eurasian Grass Snake (Family Colubridae)
Description: Size to 200 cm. Dorsum gray-green with or without black spots. Nape usually with a pair of yellow, white, orange, or red moon-shaped spots with black margins. Belly yellowish white, frequently with black spots.
Distribution and habitat: Europe, northwestern Africa, western Asia. Overgrown banks of standing or flowing water, abandoned rock quarries, gravel pits, meadows, and forests.
Care: Food: Fish, frogs, sometimes small mammals and birds, earthworms. Should be kept in a semi-moist terrarium or in an outdoor enclosure at temperatures from 20 to 30° C during the day and 15 to 20° C at night. Use a substrate of leafy forest soil and sand and add moss. Decorations should include roots and dead leaves as hiding places. Provide a large water section. The grass snake should be over-wintered in slightly damp forest soil, moss, and leaves at 5-10° C.
Behavior and breeding: If a ringed snake is threatened it will adopt a characteristic defense pose. With the help of its ribs the snake flattens its body dorso-ventrally, raises its head and neck into an S-shape, and suddenly lunges at the presumed aggressor. However, the snake does not bite but instead tries to intimidate, which usually causes an automatic withdrawal reflex on the part of the aggressor. If the snake is picked up by hand it excretes a foul-smelling substance from its post-anal glands.

The mating season for ringed snakes is in spring. A male keeps following and courting a selected female. He moves alongside her, twitching his head and body, or he slides across the female until he succeeds in embracing her tail with his own. This leads to mutual cloacal contact followed by insertion of the hemipenis.

A few weeks after copulation the female searches for a protected, damp site such as under some roots to deposit her 11-
30 eggs. The eggs have a parchment-like shell and measure
23-40 mm long with a diameter of 13-20 mm and a weight of
5-6 grams. The upper side of the eggs should be properly
marked, and they are then placed inside a plastic container
filled with damp peat moss that has been largely sterilized by
boiling. Incubation is in a Type I brood container at 28-30°
C. About 30-33 days later the hatchling will slit open its
tough egg shell with the aid of its egg tooth, which is still
solidly attached to the intermaxillary bone. Initially only the
head appears through the slit. Most hatchlings will not leave
the protective shell until a few hours later. At birth the
young are 14-22 cm long. They should be transferred into a
semi-moist terrarium and be given a diet of small fish or
commercially raised amphibian larvae. Sexual maturity is
reached at the age of 1 year.

**The following species can be kept and bred under iden-
tical conditions:**

Natrix maura
Maximum size: 100 cm.
Distribution: Southwestern Europe to northwestern Africa.
Habitat: Damp areas.
Clutch size: 32.
Incubation period and hatchling size: 40-45 days at 25-30° C;
17 cm.
Remarks: Over-winter; sexually mature in 3-4 years.

Natrix piscator—**Asian Water Snake**
Maximum size: 120 cm.
Distribution: Pakistan to Malaysia and southern Asia.
Habitat: Damp areas.
Clutch size: 87.
Incubation period and hatchling size: 84 days at 25-30° C; 19
cm.
Remarks: Over-winter.

Nerodia cyclopion—**Green Water Snake**
Maximum size: 180 cm.
Distribution: Southern USA.

Habitat: Damp areas.
Reproduction: Live-bearing; 101 young; 22 cm long.
Remarks: Over-winter.

Nerodia erythrogaster—**Plainbelly Water Snake**
Maximum size: 150 cm.
Distribution: Southern and central USA to northeastern Mexico.
Habitat: Damp areas.
Reproduction: Live-bearing; gestation 4 months; 27 young; 26 cm long.
Remarks: Over-winter.

Nerodia rhombifera—**Diamondback Water Snake**
Maximum size: 160 cm.
Distribution: South-central USA to northern Mexico.
Habitat: Damp areas.
Reproduction: Live-bearing; gestation 4 months; 62 young; 28 cm long.
Remarks: Over-winter; will also feed on turtles.

Nerodia sipedon—**Northern Water Snake**
Maximum size: 135 cm.
Distribution: Southern Canada to central USA.
Habitat: Damp areas.
Reproduction: Live-bearing; gestation 4 month; 99 young; 24 cm long.
Remarks: Over-winter.

Leimadophis poecilogyrus
Maximum size: 120 cm.
Distribution: Brazil, Ecuador, Uruguay, Argentina.
Habitat: Damp areas.
Clutch size: 7.
Incubation period: 43 days at 28-29° C.

Python regius—**Royal Python, Ball Python**
(Family Boidae)
Description: Size to 150 cm. Dorsum brown with yellowish to grayish white lateral spots. Yellow band from tip of nose to neck region. Ventral area yellow.
Distribution and habitat: Western Africa to Uganda. Open forests, savannahs.

Care: Food: Small mammals, birds. Should be kept in a semi-moist terrarium at air temperatures ranging from 26 to 32° C during the day and 22° C at night. Provide localized heating with a floor heater. Sand or a sand and peat moss mixture can serve as a substrate. Supply climbing branches and a water bowl.

Behavior and breeding: In contrast to most other snakes, this species does not defend itself by striking quickly and biting. Instead, it has developed a different defensive strategy. If the snake is threatened and there is no escape possible, it assumes a defensive posture that has given it the name "ball python." The snake coils up into a ball and hides its head among the loops. This offers the smallest possible exposed body surface for any potential enemy to attack.

According to observations in captivity, the mating season for *Pythus regius* falls into the northern spring months. The male moves alongside a female and keeps moving his tail across her back, massaging her with his erected anal spurs. If she is willing to mate she will roll her tail slightly to one side so that the male can coil his tail around hers. After a brief search for a proper hold copulation occurs. One to 4 months later the female will deposit 6 to 8 eggs that are 7-8 x 4.5-6.1 cm in size. As in many pythons, the female of *P. regius* guards her eggs. She coils her body around the clutch and places her head over it. However, as shown in recent physiological studies, there is no indication that the snake actually contributes to the correct incubation temperature, as has been shown to occur in *Python molurus*. Brood care in *P. regius,* as in most of the other pythons, is confined to selecting a suitable site for the eggs and guarding the clutch.

In order to increase the relative humidity in the terrarium, a fine mist of lukewarm water should be sprayed over the area. The incubation period for this species lasts from about 90 to 105 days. During this period the female may occasionally leave the clutch without any detrimental effect on the development of the eggs. For artificial incubation the eggs can be transferred into a Type I brood container. The substrate should consist of damp peat moss. The preferred incubation temperature falls into the range 29-33° C. Under these conditions the young will hatch after 58 days. They are 23-43 cm long and weigh about 50-53 grams. The young should be transferred to a dry terrarium. Soon after the first molt, which occurs on or about the third day, feeding will commence. The first food offered should be newly born mice.

Thamnophis sirtalis—**Common Garter Snake**
(Family Colubridae)

Description: Size to 125 cm. Dorsum black, dark brown, green, or reddish with 3 light yellow to orange bands and brown or black spots. Males with a longer tail than females.

Distribution and habitat: Southern Canada, all of USA, and into Mexico. Wetlands (swamps and around standing waters), meadows, vacant lots.

Care: Food: Fish, frogs, newts, pieces of raw fish, grasshoppers, newly born mice, thin strips of raw beef. Should be kept in a moist terrarium at air temperatures ranging from 25 to 33° C during the day and 17 to 22° C at night. Should have heating with a radiator plus regular UV radiation. The substrate should consist of a soil and peat moss mixture, plus fresh moss and leaves, to a depth of about 15 cm. Decorations can include rocks, roots, and climbing branches. A large water bowl is essential. Over-wintering in a damp leaf and peat moss mixture at about 10° C is recommended.

Behavior and breeding: Garter snakes are active during the day. When threatened they will immediately adopt a defensive posture: the body becomes flattened dorso-ventrally, the snake erects its head and neck and curls up into an S-configuration with the neck region inflated, hisses, and then suddenly lunges forward with a wide-open mouth, attempting to bite. If the snake is actually handled it will defend itself by discharging a foul-smelling secretion from its cloacal glands.

During the breeding season males are attracted to females ready to mate by an emission of a special scent given off by the skin. It is common to see several males following a particular female, flicking their tongues and crawling over her body, executing wave-like motions from head to tail. The males will press their heads against the female's back while trying to coil around her tail. If a male succeeds in establishing cloacal contact he will insert his hemipenis and copulation will occur for 15 to 20 minutes. The male may slide off the female during copulation.

Garter snakes are live-bearing. The female will produce 14 to 80 young about 4 months after the mating; the young are

18 cm long. They should be transferred to a moist terrarium and be given a diet of small fish or strips of raw fish fillets. Sexual maturity is attained after 2 years.

Care and breeding of the following species are similar to the previous species:

Thamnophis butleri—Butler's Garter snake
Maximum size: 69 cm.
Distribution: Great Lakes area, northern USA.
Habitat: Damp regions.
Litter size: 16.
Hatchling size: 15 cm.
Remarks: Over-winter; sexually mature after 2 years.

Thamnophis cyrtopsis—Blackneck Garter Snake
Maximum size: 110 cm.
Distribution: Southwestern USA.
Habitat: In the proximity of water; deserts to forests.
Litter size: 25.
Juvenile size: 23 cm.
Remarks: Over-winter.

Thamnophis elegans—Western Terrestrial Garter Snake
Maximum size: 110 cm.
Distribution: Western USA to Mexico.
Habitat: In the· proximity of water.
Litter size: 19.
Juvenile size: 19 cm.
Remarks: Over-winter; sexually mature after 1 year.

Opheodrys aestivus—Rough Green Snake
Maximum size: 120 cm.
Distribution: Central and eastern USA.
Habitat: Among vegetation, often close to water.
Clutch size: 12.
Incubation period and hatchling size: 42 days at 20-24° C; 17 cm.
Remarks: Over-winter.

LITERATURE

Anonymus (1974) Amphibians. A report. National Academy of Sciences. Washington D. C.

ARNOLD, E. N., J. A. BURTON (1979) Pareys Reptilien- und Amphibienführer Europas. Verlag Paul Parey. Hamburg, Berlin.

ARNOLD, S. J. (1976) Sexual behavior, sexual interference and sexual defense in the salamanders *A. maculatum, A. tigrinum and P. jordani.* Z. Tierpsychol. **42.**

AUFFENBERG, W. (1977) Display behavior in tortoises. Amer. Zool. **17.**

BARASH, D. P. (1980) Soziobiologie und Verhalten. Verlag Paul Parey. Hamburg, Berlin.

BEHLER, J. L., F. WAYNE (1979) The audubon society field guide to north american reptiles an amphibians. A. A. Knopf. New York.

BEHRMANN, H.-J. (1981) Haltung und Nachzucht von *Varanus timorensis t.* Salamandra **17.**

BEUTLER, A., U. GRUBER (1978) Geschlechtsdimorphismus, Populationsdynamik und Ökologie von *Cyrtodactylus kotschyi.* Salamandra **15.**

BISCHOFF, W. (1974) Beobachtungen bei der Pflege von *Lacerta simonyi stehlinii.* Salamandra **10.**

BISCHOFF, W. (1981) Freiland- und Terrarienbeobachtungen an der Omaneidechse *Lacerta jayakari.* Z. Kölner Zoo **24.**

BLASCO, M. (1979) *Chamaeleo chamaeleon* in the province of Malaga, Spain. Brit. J. Herp. **5.**

BÖHME, W. (1975) Indizien für natürliche Parthenogenese beim Helmbasilisken. Salamandra **11.**

BÖHME, W. (1978) Zur Herpetofaunistik des Senegal. Bonn. Zool. Beitr. **29.**

BÖHME, W.; Hrsg. (1981) Handbuch der Reptilien und Amphibien Europas (Bd. 1). Akad. Verlagsgesellschaft. Wiesbaden.

BÖHME, W., W. BISCHOFF (1976) Paarungsverhalten der kanarischen Eidechsen als systematisches Merkmal. Salamandra **12.**

BRAUNWALDER, M. E. (1979) Über eine erfolgreiche Zeitigung von Eiern des Grünen Leguan und die damit verbundene Problematik. Salamandra **13.**

BREHM, A. (1869) Illustriertes Tierleben. 5. Bd.

Kriechtiere, Lurche, Fische. Verlag des Bibliogr. Instituts, Hildburghausen.

BRÖER, W. (1981) Über die erfolgreiche Nachzucht einer F_2-Generation der Rotschwanznatter. Salamandra **17.**

BROCKHUSEN, F. v. (1977) Untersuchungen zur individuellen Variabilität der Beutenahme von *Anolis lineatopus.* Z. Tierpsychol. **44.**

BRUNO, S., S. MAUGERI (1976, 1978) Rettili d'italia. Vol. 1, 2. Aido Martello. Guinti Editore S.P.A.

BUDDE, H. (1980) Verbesserter Brutbehälter zur Zeitigung von Schildkrötengelegen. Salamandra **16.**

BUNNELL, P. (1973) Vocalizations in the territorial behavior of the frog *D. pumilio.* Copeia.

BURGHARDT, G. M. (1977) Neonate reptile behavior. Amer. Zool. **17.**

BUSTARD, H. R. (1965) Observations on the life history and behavior of *Chamaeleo hohnelii.* Copeia **4.**

BUSTARD, H. R. (1967) The comparative behavior of chamaeleons. Herpetologica **23.**

BUSTARD, H. R. (1970) Australian Lizards. Collins. Sydney, London.

BUSTARD, H. R. (1971) A population study of the eyed gecko *Oedura ocellata* in northern New South Wales, Australia. Copeia **4.**

CARPENTER, C. (1960) Parturition and behavior of birth at *Sceloporus jarrovi.* Herpetologica **16.**

CHIRAS, S. (1982) Captive reproduction of the children's python. Herp. Review **13.**

CLYNE, D. (1969) Australian frogs. Landsdowne Press. Melbourne.

COCHRAN, D. M. (1970) Knaurs Tierreich in Farben, Amphibien, Droemersche Verlagsanstalt, Th. Knaur. Kremyr u. Scheriden. Wien.

COE, M. (1974) Observations on the ecology and breeding of the Genus *Chiromantes.* J. Zool. Lond. **172.**

COGGER, H. G. (1979) Reptiles and amphibians of Australia. A. H. u. A. W. Reed. Pty. Ltd. Sydney.

COOPER, W. E. (1979) Variability and predictability of courtship in *Anolis carolinensis.* J. Herpetol **13.**

LITERATURE

CRESPO, E. G. (1973) Sobre a distribuçao e ecologia da herpetofauna portuguesa. Arq. Mus. Bocage. Lisboa 4.

CURIO, E., H. MÖBIUS (1978) Versuch zum Nachweis des Riechvermögens bei *Anolis lineatopus*. Z. Tierpsychol. 47.

DECKERT, K. (1961) Die Paarung des Beutelfrosches. Zoolog. Garten. NF.

DUELLMAN, W. E. (1966) Aggressive behavior in dendrobatid frogs. Herpetologica 22.

DUELLMAN, W. E. (1970) Hylid frogs of Middle America, Vol 1,2. Monograph of the Museum of National History. The University of Kansas.

DUELLMAN, W. E. (1978) The biology of an equatorial herpetofauna in Amazonian Ecuador. Univ. Kans. Lawrence. Miscell. Public, No. 65.

EHRENGART, W. (1971) Zur Pflege und Zucht der Griechischen Landschildkröte. Salamandra 7.

EHRENGART, W. (1976) Brutanlagen für Schildkröteneier. Salamandra 12.

EIBL-EIBESFELDT, J. (1950) Ein Beitrag zur Paarungsbiologie der Erdkröte. Behaviour 4.

EIBL-EIBESFELDT, J. (1951) Nahrungserwerb und Beuteschema der Erdkröte. Behaviour 4.

EIBL-EIBESFELDT, J. (1952) Vergleichende Verhaltensstudien an Anuren. Z. Tierpsychol. 4.

EIBL-EIBESFELDT, J. (1956) Vergleichende Verhaltensstudien an Anuren. Zool. Anz. Suppl. 19.

EIBL-EIBESFELDT, J. (1978) Grundriß der vergleichenden Verhaltensforschung. R. Piper & Co. Verlag. München, Zürich.

EIKHORST, R., W. EIKHORST (1982) Zur Fortpflanzung der Spanischen Kieleidechse. Salamandra 18.

ELZEN, P. v. d. (1979) Remarques sur *Bombina orientalis*. Rev. fr. Aquariol. 6.

ENSINCK, F. H. (1980) De kweek von *Dendrobates tinctorius*. Lacerta 10–11.

ERNST, C. H. (1981) Courtship behavior of male *Terrapene carolina major*. Herp. Review 12.

EWERT, J. P. (1976) Neuro-Ethologie. Springer Verlag. Berlin, Heidelberg, New York.

FAUCI, J. (1981) Breeding and rearing of captive solomon island ground boas. Herp. Review 12.

FILEK, W. v. (1967) Frösche im Aquarium. Franckh'sche Verlagshandlung. Stuttgart.

FITCH, H. S. (1970) Reproductive cycles in lizards and snakes. Mus. Nat. Hist. Kansas.

FRANK, W. (1976) Parasitologie. Verlag E. Ulmer. Stuttgart.

FRANK, W. (1978) Schlangen im Terrarium. Franckh'sche Verlagshandlung. Stuttgart.

FREYTAG, G. E. (1970) Beobachtungen zum Paarungsverhalten von *Cynops pyrrhogaster*. Salamandra 6.

FREYTAG, G. E., G. PETERS et. al. (1973) Urania Tierreich. Fische, Lurche, Kriechtiere. Urania Verlag. Leipzig, Jena, Berlin.

FRIEDERICH, U., W. VOLLAND (1981) Futtertierzucht. Verlag Eugen Ulmer. Stuttgart.

GANS, C., S. A. BELLAIRS, T. S. PARSONS, Hrsg. (1969–1980) Biology of Reptilia. Vol. 1–10. Academic Press. London, New York.

GLÄSS, H., W. MEUSEL (1969) Die Süßwasserschildkröten Europas. A. Ziemsen Verlag. Wittenberg, Lutherstadt.

GOLDER, F. (1972) Beitrag zur Fortpflanzungsbiologie einiger Nattern. Salamandra 8.

GOLDER, F. (1981) Anomalien bei der Fortpflanzung von *Elaphe guttata* g. Salamandra 17.

GRAF, A. B. (1980) Exotica. Pictoal Cyclopedia of Exotic Plants. Roehrs Company Inc. E. Rutherford. New York.

GREEN, H. W. (1970) Beobachtungen zur Fortpflanzung von *Sceloporus poinsetti*. Salamandra 6.

GREENBERG, N. (1977) Anolis neuroethology. Amer. Zool. 17.

GREVEN, H. (1976) Geburtsvorgang beim Feuersalamander. Salamandra 12.

GREVEN, H. (1977) Comparative ultrastructural investigations of the uterine epithelium in the viviparous *Salamandra atra* and the ovoviviparous *S. salamandra*. Cell. Tiss. Res. 181.

GRIEHL, K. (1982) Schlangen. Gräfe und Unzer. München.

GRZIMEKS Tierleben (1970, 1971) Bd. V: Fische 2 und Lurche. Bd. VI: Kriechtiere. Kindler Verlag. Zürich.

HAGDORN, H. (1973) Beobachtungen zum Verhalten von Phelsumen. Salamandra 9.

HALLIDAY, T. H. (1975) An observational and experimental study of sexual behavior in the smooth newt *T. vulgaris*. Anim. Behav. 23.

HELVERSEN, O. v. (1979) Angeborenes Erkennen akustischer Schlüsselreize. Verh. Dtsch. Zool. Ges.

LITERATURE

HEMMER, H. (1977) Studien an einer nordwest-deutschen Grünfroschpopulation als Beitrag zur Bestimmungsproblematik und zur Rolle der Selektion im *Rana esculenta*-Komplex. Salamandra 13.

HEUSSER, H. (1958, 1960) Über die Beziehungen der Erdkröte zu ihrem Laichplatz. Behaviour XII, XVI.

HIMSTEDT, A. (1969) Über *Sceloporus jarrovi*. Salamandra 5.

HIMSTEDT, W., C. PLASA, K. HELLER (1979) Erkennen stationärer visueller Muster. Verh. Dtsch. Zool. Ges.

HONEGGER, R. (1970) Beitrag zur Fortpflanzungsbiologie von *Boa constrictor* und *Python reticulatus*. Salamandra 6.

HOOGMOED, M. S. (1969) Notes on the herpetofauna of Surinam. Zoolog. Meded. 44.

HORN, H. G. (1978) Nachzucht von *Varanus gilleni*. Salamandra 14.

HORN, H. G., G. PETERS (1982) Beiträge zur Biologie des Rauhnackenwarans, *Varanus rudicollis*. Salamandra 18.

HUN, E. (1972) Erfolgreiche Zucht des Grünen Leguans. Salamandra 8.

JAHN, J. Lebendfutter für ausgewachsene Aquarien- und Terrarientiere. Albrecht Philler Verlag. Minden/Westf.

JENSSEN, T. H. (1977) Evolution of anoline display. Amer. Zool. 17.

JOCHER, W. (1969) Schildkröten. Franckh'sche Verlagshandlung. Stuttgart.

JOHNSON, J. A., E. D. BROODIE (1974) Defensive behavior of the western banded gecko *Coleonyx variegatus*. Anim. Behav. 22.

KABISCH, K. (1978) Die Ringelnatter. Neue Brehm Bücherei. A. Ziemsen Verlag. Wittenberg, Lutherstadt.

KÄSTLE, W. (1964) Verhaltensstudien an Taggeckonen. Z. Tierpsychol. 21.

KÄSTLE, W. (1967) Soziale Verhaltensweisen von Chamäleons der *pumilus*- und *bitaeniatus*-Gruppe. Z. Tierpsychol. 24.

KÄSTLE, W. (1979) Echsen im Terrarium. Franckh'sche Verlagshandlung. Stuttgart.

KIESTER, A. M. (1977) Communication in amphibians and reptiles. In: How animals communicate. (Th. A. Seboek, Hrsg.) Indiana Univ. Press. Bloomington, London.

KITZLER, G. (1941) Die Paarungsbiologie einiger Eidechsen. Z. Tierpsychol. 4.

KLAGES, H. G. (1982) Pflege und Nachzucht der australischen Bodenagame *Amphibolurus nuchalis*. Salamandra 18.

KLEMMER, K. (1976) The amphibia and reptilia of the canary islands. In: Biogeography and ecology in the canary islands. (Kunkel, G.; Hrsg.). Dr. W. Junk B. V. Publishers.

KLINGELHÖFFER, W. (1955, 1956, 1957, 1959) Terrarienkunde. Bd. 1-4. Alfred Kernen Verlag. Stuttgart.

KLÖS, H.-G., E. M. LANG; Hrsg. (1976) Zootierkrankheiten. Verlag Paul Parey. Berlin, Hamburg.

LAMOTTE, M., J. LESCURE (1977) Tendences adaptives à l'affranchissement du milieu aquatique chez les amphibiens anoures. Ter. et Vie 2.

LANGERWERF, B. (1980) The armenian wall lizard *Lacerta armeniaca* with notes on its care and reproduction in captivity. Brit. Herp. Soc. Bull. 2.

LANGERWERF, B. (1981) Agama stellio with observations on its care and reproduction in captivity. Brit. Herp. Soc. Bull. 2.

LANGERWERF, B. (1981) The southern alligator lizard *Gerrhonotus multicarinatus*: Its care and breeding. Brit. Herp. Soc. Bull. 4.

LEHRMANN, H. P. (1970) Beobachtungen bei der Haltung und Aufzucht von *Trachyboa boulengeri*. Salamandra 6.

LEVITON, A. Reptiles and amphibians of North America. Doubleday u. Company Inc. New York.

LIEBERMANN, A. (1980) Nesting of the basilisk lizard. J. Herpetol. 14.

LILGE, D., H. V. MEUWEN (1979) Grundlagen der Terrarienhaltung. Landbuch Verlag. Hannover.

LIN, J., C. E. NELSON (1981) Comparative reproductive biology of two sympatric tropical lizards. Amphibia-Reptilia 1.

LÜDDECKE, H. (1974) Ethologische Untersuchungen zur Fortpflanzung von *Phyllobates palmatus*. Dissertation, Mainz.

MARCELLINI, D. (1977) Display behavior in gekkonid lizards. Amer. Zool. 17.

MATUSCHKA, F.-R. (1978) Beobachtungen bei der Haltung von *Otocryptis wiegmanni*. Salamandra 14.

MATZ, G., M. VANDERHAEGE (1980) BLV Terrarienführer. BLV Verlagsgesellschaft. München, Bern, Wien.

MAU, K.-G. (1978) Nachweis natürlicher Par-

374

LITERATURE

thenogenese bei *Lepidodactylus lugubris* durch Gefangenschaftsnachzucht. Salamandra **14.**

MEBS, D. (1966) Studien zum aposemantischen Verhalten von *Teratoscincus scincus.* Salamandra **2.**

MEBS, D. (1975) Herpetologische Beobachtungen in NSW-Australien. Salamandra **11.**

MEEDE, U. (1980) Beobachtungen an *Dendrobates quinquevittatus* und *Phyllobates femoralis.* Salamandra **16.**

MEIER, A. (1977) Beobachtungen an *Phelsuma standingii.* Salamandra **13.**

MEMBERS, A. H. S. (1976) Observations on the easter water dragon. Herpetofauna **8.**

MENZIES, J. J. (1957) Breeding behaviour of the chamaeleon *Chamaeleo gracilis.* Brit. J. Herp.

MERTENS, R. (1929) Aus dem Leben eines Faltengeckos. Natur und Museum. Heft **4.**

MERTENS, R. (1946) Die Warn- und Drohreaktionen der Reptilien. Abh. Senckenberg. Naturf. Gesell. **471.**

MERTENS, R. (1968) Kriechtiere und Lurche. Franckh'sche Verlagshandlung. Stuttgart.

MERTENS, R., H. WERMUTH (1960) Die Amphibien und Reptilien Europas. Verlag Waldemar Kramer. Frankfurt a. M.

MIEROP, L. H. S. v., E. L. BESETTE (1981) Reproduction of the ball python in captivity. Herp. Review **12.**

MUDRACK, W. (1972) Haltung und Zucht der Chinesischen Rotbauchunke *Bombina orientalis.* Salamandra **8.**

MURPHY, J. B., C. A. MITCHELL (1974) Ritualized combat behavior of the pygmy mulga monitor lizard *Varanus gilleni.* Herpetologica **30.**

MURPHY, J. B., J. T. COLLINS (1980) Reproductive biology and diseases of captive reptiles. Contr. Herp. **1.** SSAR.

MYERS, C. M. (1982) Spotted poison frogs. Am. Mus. Novit. **2721.**

MYERS, C. M., J. W. DALY (1976) Preliminary evaluation of skin toxins and vocalizations in taxonomic and evolutionary studies of poison-dart frogs. Bull. Amer. Mus. Nat. Hist. Vol **157** Art. 3. New York.

MYERS, C. M., J. W. DALY (1979) A name for the poison frog of Cordillera Azul, Eastern Peru. Am. Mus. Novit. **2674.**

MYERS, C. M., J. W. DALY, B. MALKIN (1978) A dangerously toxic new frog used by embera in-dians of Western Colombia. Bull. Am. Mus. Nat. Hist. **161.**

NETTMANN, H. K., S. RYKENA (1979) Mauergeckos, die ihre Eier im Sand vergraben. Salamandra **15.**

NIEKISCH, M. (1975) Pflege und Zucht von *Egernia cunninghami.* Salamandra **11.**

NIEKISCH, M. (1980) Terraristische Beobachtungen zur Biologie von *E. cunninghami.* Salamandra **16.**

NIETZKE, G. (1978) Die Terrarientiere 1 und 2. Verlag Eugen Ulmer. Stuttgart.

NOLAN, M. (1981) Notes on the care and captive breeding of the sinaloan milk snake. Brit. Herp. Soc. Bull. **4.**

OBST, F. J. (1980) Schildkröten. Neumann Verlag. Leipzig, Jena, Berlin.

ORTLEB, E. (1965) Hatching of basilisk eggs. Herpetologica **20.**

ORTLEB, E. (1972) Observations of fish-eating and maintenance behavior in two species of Basiliscus. Copeia **2.**

PARKER, H. W., A. BELLAIRS (1971) Les amphibiens et les reptiles. Bordas. Paris, Montreal.

PASSMORE, N. I., V. C. CORRUTHEN (1979) South african frogs. Witwatersrand Univ. Press. Johannesburg.

PERRON, U. (1974) Haltung und Nachzucht von *Basiliscus b.* und *Basiliscus plumifrons.* Salamandra **10.**

PETZOLD, H.-G. (1971) Blindschleiche und Scheltopusik. A. Ziemsen Verlag. Wittenberg, Lutherstadt.

PETZOLD, H.-G. et. al. (1979) Wildtiere in Menschenhand Bd. 1. VEB Deutscher Landwirtschaftsverlag, Berlin.

PLASA, C. (1979) Heimfindeverhalten bei *Salamandra s.* Z. Tierpsychol. **51.**

PRECHTL, H. F. R. (1951) Zur Paarungsbiologie einiger Molcharten. Z. Tierpsychol. **8.**

RAAB, G. B., M. S. RAAB (1963) On the behavior and breeding biology of the African Pipid frog. Z. Tierpsychol. **20.**

RAEHMEL, Ch. A. (1974) Eine gelungene Kreuzung zwischen *Bombina variegata v.* und *Bombina orientalis.* Salamandra **10.**

RAUH, W. (1969, 1973) Bromelien für Zimmer und Gewächshaus. Eugen Ulmer. Stuttgart.

REICHENBACH-KLINKE, H. H. (1961) Krankhei-

LITERATURE

ten der Amphibien. Gustav Fischer Verlag. Stuttgart.

REICHENBACH-KLINKE, H. H. (1963) Krankheiten der Reptilien. Gustav Fischer Verlag. Stuttgart.

RENSCH, B., Ch. ADRIAN-HINSBERG (1963) Die visuelle Lernkapazität von Leguanen. Z. Tierpsychol. **20.**

RICHTER, W. (1974) Orchideen pflegen, vermehren, züchten. Verlag Neumann-Neudamm. Melsungen.

RIMPP, K. (1978) Molche und Salamander. Verlag Eugen Ulmer. Stuttgart.

ROTTER, J. (1963) Die Warane. A. Ziemsen Verlag. Wittenberg, Lutherstadt.

RYKENA, S., H.-K. NETTMANN (1974) Eine gelungene Kreuzung von *Lacerta trilineata* und *Lacerta viridis*. Salamandra **10.**

RYKENA, S., C. HENKE (1978) Bastardierung von *Lacerta agilis* × *Lacerta viridis*. Salamandra **14.**

SALVADOR, A. (1974) Guia de los anfibios y reptiles españoles. Mejorada del Campo. Madrid.

SACHSSE, W. (1975) Jährliche Nachzucht bei der Chinesischen Dreikielschildkröte. Salamandra **11.**

SACHSSE, W. (1976) Nachzucht in der 2. Generation von *Staurotypus salvinii*. Salamandra **12.**

SACHSSE, W. (1977) Normale und pathologische Phänomene bei Zuchtversuchen mit Schildkröten. Salamandra **13.**

SACHSSE, W. (1977) *Sternotherus minor m.*, seine Nachzucht und die damit verbundenen biologischen Beobachtungen. Salamandra **13.**

SACHSSE, W. (1980) Zur Biologie von *Kinosternon leucostomum* in Gefangenschaft I. Salamandra **16.**

SACHSSE, W. (1980) Zur Biologie und Fortpflanzung von *Kinixys belliana nogueyi*. Salamandra **16.**

SAVAGE, J. M. (1973) The geographic distribution of frogs. Spec. Scien. Contr. Allan H. Foundation.

SAVAGE, J. M. (1980) A handlist with preliminary keys to the herpetofauna of Costa Rica. Allan H. Foundation.

SCHIØTZ, A. (1975) The treefrogs of eastern Africa. Steen strupia. Copenhagen.

SCHLEICH, H.-H. (1979) Feldherpetologische Beobachtungen in Persien. Salamandra **15.**

SCHMIDT, A. A. (1970) Zur Fortpflanzung der Kreuzbrustschildkröte. Salamandra **6.**

SCHMIDT, A. A. (1976) Nachzucht von *Bufo blombergi*. Salamandra **12.**

SCHMIDT, A. A. (1976) Erstnachzucht des Zipfelfrosches. Salamandra **12.**

SCHMIDT, A. A. (1978) Erstnachzucht des Indischen Ochsenfroschs. Salamandra **14.**

SCHMIDT, A. A., R. WICKER (1977) Weitere Beobachtungen bei der Nachzucht des Zipfelfroschs. Salamandra **13.**

SCHMIDT, K. P., R. F. INGER (1969) Knaurs Tierreich in Farben. Reptilien. Droemersche Verlagsanstalt. Th. Knaur. Wien.

SCHMIDT, D. (1979) Schlangen im Terrarium. Neumann Verlag. Leipzig.

SCHMIDT, D. (1981) Echsen im Terrarium. Neumann Verlag. Leipzig.

SCHNEIDER, F. Die Pflanzen des Terrariums. Albrecht Philler Verlag. München.

SCHNEIDER, H. (1966) Die Paarungsrufe einheimischer Froschlurche. Z. Morph. Ökol. Tiere **57.**

SCHNEIDER, H. (1981) Fortpflanzungsverhalten des Mittelmeerlaubfrosches. Salamandra **17.**

SCHULTE, R. (1980) Frösche und Kröten. Verlag Eugen Ulmer. Stuttgart.

SCHUSTER, B. (1980) Zur Biologie von *Kinosternon leucostomum* in Gefangenschaft II. Salamandra **16.**

SCHUSTER, M. (1979) Experimentelle Untersuchungen zum Beutefang-, Kampf- und Fortpflanzungsverhalten von *Chamaeleo jacksonii*. Dissertation Münster.

SENFFT, W. (1936) Das Brutgeschäft des Baumsteigerfrosches *D. auratus*. D. Zool. Garten **8.**

SILVERSTONE, P. A. (1973) Observations on the behavior and ecology of a colombian poison arrow frog. Herpetologica **29.**

SILVERSTONE, P. A. (1975) A revision of the poison-arrow frogs of the genus *Dendrobates*. Nat. Hist. Mus. Los Angeles County. Sci. Bull. **21.**

SILVERSTONE, P. A. (1976) A revision of the poison-arrow frogs of the genus *Phyllobates*. Nat. Hist. Mus. Los Angeles County. Sci. Bull. **27.**

SIMON, C. A. (1981) The role of chemoreception in the iguanid lizard *Sceloporus jarrovi*. Anim. Behav. **29.**

SMITH, R. J. (1975) Breeding and rearing *Bufo*

LITERATURE

blombergi. Int. Zoo Yb. **15.**

STEMMLER, C. (1969) Paarungsverhalten von *Eryx johnii*. Salamandra **5.**

STEPHENSON, G. (1977) Notes on *Tiliqua gerardi* in captivity. Herpetofauna **9.**

STETTLER, P. H. (1981) Handbuch der Terrarienkunde. Franckh'sche Verlagshandlung. Stuttgart.

STEWART, M. W. (1967) Amphibians of Malawi. State University. New York.

STIRNBERG, E., H. G. HORN (1981) Eine unerwartete Nachzucht im Terrarium. Varanus (Odatria) storri. Salamandra **17.**

TARDENT, D. (1972) Haltung und Zucht von Sternschildkröten. Salamandra **8.**

TAYLOR, E. H. (1952) A review of frogs and toads of Costa Rica. Sci. Bull. Kan. Vol. XXXVIII.

TAYLOR, E. H. (1963) The lizards of Thailand. Sci. Bull. **14.**

TAYLOR, D. H., S. I. GUTTMAN; Hrsg. (1977) The reproductive biology of amphibians. Plenum Press. N. Y., London.

TORKARZ, R. R., R. C. JONES (1979) A study on egg-related maternal behavior in *Anolis carolinensis*. J. Herpetol. **13.**

TRUTNAU, L. (1975) Europäische Amphibien und Reptilien. Belser Verlag.

TRUTNAU, L. (1979) Schlangen Bd. 1. Verlag Eugen Ulmer. Stuttgart.

VERBEEK, B. (1972) Über die Haltung und Zucht von *Lacerta hispanica*. Salamandra **8.**

VOGT, P. (1974) Breeding and rearing the colombian giant toad. Int. Zoo Yb. **14.**

VOKINS, M. (1977) Breeding of *Geochelone carbonaria*. J. Jersey Wildlife Preservation Trust, 14.

WAGER, V. A. (1965) The frogs of South Africa. Purnell u. Sons. Pty. London, Cape Town, Johannesburg.

WAGNER, E. (1979) Breeding king snakes. Int. Zoo Yb. **19.**

WAHLERT, W. v. (1965) Molche und Salamander. Franckh'sche Verlagshandlung. Stuttgart.

WASSERSUG, R. J., K. J. FROGNER, R. F. INGER (1981) Adaptions for the life in tree holes by rhacophorid tadpoles from Thailand. J. Herpetol. **15.**

WEBER, H. (1957) Vergleichende Untersuchungen des Verhaltens von Smaragdeidechsen. Z. Tierpsychol. **14.**

WELLS, K. D. (1977) The social behavior of anuran amphibians. Anim. Behav. **25.**

WELLS, K. D. (1980) Social behavior and communication of a dendrobatid frog, *Colostethus trinitatis*. Herpetologica **36.**

WELLS, K. D. (1980) Behavioral ecology and social organization of a dendrobatid frog, *Colostethus inguinalis*. Behav. Ecol. Sociobiol. **6.**

WELLS, K. D. (1981) Parental behavior of male and female frogs. In: Natural selection and social behavior. Recent research and new theory. (ALEXANDER, R. D.; TINKLE, D. W.; Hrsg.). Chiron Press. New York.

WERMUTH, H., R. MERTENS (1961) Schildkröten, Krokodile, Brückenechsen. VEB Gustav Fischer Verlag. Jena.

WERNER, Y. L. (1965) Über die israelischen Geckos der Gattung *Ptyodactylus*. Salamandra **1.**

WERNER, Y. L., E. A. FRANKENBERG (1978) Further observations on the vocal repertoire of *Ptyodactylus hasselquistii*. Isr. J. Zool. **27.**

WEYGOLDT, P. (1976) Beobachtungen zur Biologie und Ethologie von *Pipa carvalhoi*. Z. Tierpsychol. **40.**

WEYGOLDT, P. (1980) Zur Fortpflanzungsbiologie von *Phyllobates femoralis* im Terrarium. Salamandra **16.**

WEYGOLDT, P. (1981) Complex brood care and reproductive behavior in captive poison-arrow frogs. *Dendrobates pumilio*. Behav. Ecol. Sociobiol. **7.**

WICKLER, W., U. SEIBT (1974) Rufe und Antworten bei *Kassina senegalensis, Bufo regularis* und anderen Anuren. Z. Tierpsychol. **34.**

WILKE, H. (1979) Schildkröten. Gräfe und Unzer Verlag. München.

ZIMMERMANN, E., H. ZIMMERMANN (1982) Soziale Interaktionen, Brutpflege und Zucht des Pfeilgiftfrosches *Dendrobates histrionicus*. Salamandra **18.**

ZIMMERMANN, E., P. ZIMMERMANN, A. ZIMMERMANN (1975) Eine ökologische Untersuchung im Lebensraum des gemeinen Chamäleons in SW-Spanien. MS.

ZIMMERMANN, H. (1978) Tropische Frösche. Franckh'sche Verlagshandlung. Stuttgart.

ZIMMERMANN, H. (1982) Futtertier-ABC. Franckh'sche Verlagshandlung. Stuttgart.

ZIMMERMANN, H., E. ZIMMERMANN (1981) Sozialverhalten, Fortpflanzungsverhalten und Zucht einiger Dendrobatiden. Z. Kölner Zoo **21.**

Index

378